◎ 图联社 编著

黄河口国家公园

100 个
决定性瞬间

The Yellow River Estuary National Park

100 Decisive

MOMENTS

山東畫報出版社
济南

图书在版编目（CIP）数据

黄河口国家公园100个决定性瞬间 / 图联社编著.
济南 : 山东画报出版社, 2024. 11. -- ISBN 978-7
-5474-5190-8

Ⅰ. S759.992.52-64

中国国家版本馆CIP数据核字第20241J3Y86号

HUANGHEKOU GUOJIA GONGYUAN 100 GE JUEDINGXING SHUNJIAN

黄河口国家公园 100 个决定性瞬间

图联社 编著

选题策划 姜 辉
责任编辑 马 赛
装帧设计 圖聯社 TLS

主管单位 山东出版传媒股份有限公司
出版发行 山东画报出版社

社　　址 济南市市中区舜耕路 517 号　邮编 250003
电　　话 总编室（0531）82098472
市场部（0531）82098479　82098476（传真）
网　　址 http://www.hbcbs.com.cn
电子信箱 hbcb@sdpress.com.cn

印　　刷 济南新先锋彩印有限公司
规　　格 210 毫米 ×260 毫米　16 开
12 印张　50 千字
版　　次 2024 年 11 月第 1 版
印　　次 2024 年 11 月第 1 次印刷
书　　号 ISBN 978-7-5474-5190-8
定　　价 298.00 元

– 特别鸣谢 –

为本书提供影像作品的
全体摄影师们

Special Thanks to All Photographers
Who Contributed Their Images to This Book

PREFACE

黄河入海
万鸟齐飞

张劲硕

中国科学院动物研究所博士
国家动物博物馆馆长、研究员

“白日依山尽，黄河入海流。欲穷千里目，更上一层楼。”这首国人从小就背诵的唐代诗人王之涣之名篇《登鹳雀楼》，可以说，是我们儿时接触的最早为我们科普“黄河入海”地理知识的“文章”了！

然而，鹳雀楼位于山西省永济市，蒲州古城之西，黄河之东。我怀疑王之涣老先生未必莅临过“黄河入海”之处。若来到山东省东营市，这片黄河入海之地，观其景，察其色，望其水，赏其鸟，王老先生又会有怎样的感慨和诗篇呢？！

我虽生长在北京市，但祖籍是山东省烟台市福山区，我的家乡与东营市隔海而望，这里有我国暖温带最完整的河口湿地生态系统。说到这里，至今仍有许多人迷惑，“湿地”不就是过去的沼泽、烂滩、水泡子吗？！ 20 世纪六七十年代，我们曾一度提出“向沼泽进军”“向沼泽要粮”“放干沼泽”的口号，——甚至当时还没有“湿地”（wetlands）的科学概念。因为当时我们对生态系统缺乏科学认识，甚至以为那里是“荒地”，需要我们去改造。许多年后的今天，我们已然认识到，那不是“荒地”，而是一片广袤的“荒野”，是人类的“自然银行”。

1992 年，我国正式加入《关于特别是作为水禽栖息地的国际重要湿地公约》，简称《湿地公约》，亦称《拉姆萨公约》。从此，开启了我国研究和保护湿地之历程。三十多年来，我国生态科学、湿地研究得到长足发展，生态文明建设已经成为国策、写入《宪法》，人们的生态保护意识不断提高。湿地作为一种极为重要、不可或缺的生态系统类型，得到包括国人在内的全世界人民的广泛认可和保护。2024 年 7 月，中国黄（渤）海候鸟栖息地（第二期）被列入《世界遗产名录》，成为全世界最重要的自然遗产之一，在全球湿地生态系统保护中得到承认和关注。

从另外一个维度讲，2013 年，党的十八届三中全会首次提出建立国家公园体制，并将其列入全面深化改革的重点任务，中国特色国家公园体制建设正式起步。2021 年 10 月，中国正式设立三江源、大熊猫、东北虎豹、海南热带雨林、武夷山等第一批 5 个国家公园，保护面积达 23 万平方公里，涵盖近 30% 的陆域国家重点保护野生动植物种类。至此，我国的“以国家公园为主体的自然保护地体系”建设得到全面提升与完善，也引领从国家到地方各个层面积极推进国家公园建设。其中，正在规划建设的“黄河口国家公园”是我们最为期盼的国家公园之一！

据我了解，黄河口国家公园将整合 8 个自然保护地，总面积达 3518 平方公里，包括陆域 1371 平方公里，海域 2147 平方公里。从这一规模上看，我们不难理解，黄河口国家公园一旦建成，将更为完整地保护好黄河河口湿地生态系统。而且，最为难能可贵的是，这片湿地是我国沿海地区人口最为稀少的

序言

区域，意味着包括鸟类在内的各类野生动物受到人为干扰也是最少的。

近些年，我们中国科学院动物研究所，以及兄弟单位如北京师范大学、中国环境科学学院等诸多生物学、生态学、环境科学研究单位会同当地保护地、野生动物主管部门，与当地各级政府、机构深入合作，对黄河口一带的自然地理、生态环境、野生动植物多样性等方方面面做了系统而深度的研究。科学家们一致认为，黄河口国家公园是国际候鸟迁徙关键区域，更是渤海、黄海水生生物种质资源库！

所以，我个人最企盼的便是，未来不仅仅是本书这些优秀的摄影师来黄河口拍摄旖旎风光、百鸟千姿；更有价值和意义的是，让社会公众，特别是青少年，可以走进黄河口国家公园领略自然遗产地之风貌，感受生灵之美，丰富人们的精神生活！

“云销雨霁，彩彻区明。落霞与孤鹜齐飞，秋水共长天一色。”我们在欣赏这部精美画册的同时，一定要为自己规划一次旅程，有机会来到东营，来到黄河口国家公园，身临其境地感受一下荒野自然之壮美，以及飞羽之优雅。

阅此书，行此路！是为序。

2024 年 11 月 1 日

于国家动物博物馆

The Yellow River flows into the sea and thousands of birds fly together

Zhang Jinshuo

Director, National Zoological Museum of China
Professor, Institute of Zoology, Chinese Academy of Sciences

"The sun along the mountain bows, the Yellow River seaward flows. You will enjoy a grander sight, if you climb to a greater height". This famous poem,"On the Stork Tower"by Wang Zhihuan, a Tang dynasty poet, is something that Chinese people have recited since childhood. It can be said to be one of the earliest works that introduces us to the geographical knowledge of the"Yellow River entering the sea".

However, the Stork Tower is located in Yongji City, Shanxi Province, west of the ancient city of Puzhou and east of the Yellow River. I suspect that Mr. Wang may not have visited the actual point where the Yellow River meets the sea. If he were to come to Dongying City in Shandong Province, the place where the Yellow River flows into the sea, he could observe its scenery, perceive its colors, gaze at its waters, and admire its birds. Then, what sentiments and poems might he express?

Though I grew up in Beijing, my ancestral home is in Fushan District, Yantai City, Shandong Province, and my hometown looks across the sea to Dongying City. Here lies the most complete estuarine wetland ecosystem in China's warm temperate zone. Speaking of this, many people are still puzzled: Isn't a "wetland" just a former marsh, mudflat, or puddle? In the 1960s and 1970s, we even proposed slogans like "March into the marshes", "Get grain from the marshes", and "Drain the swamps completely"—at that time, the scientific concept of "wetlands" had not even been established. Our lack of scientific understanding of ecosystems led us to mistakenly believe that these were "wastelands" needing our transformation. Many years later, we have come to recognize that it is not "wasteland", but a vast "wilderness", serving as humanity's "natural bank".

In 1992, China officially joined the "Convention on Wetlands of International Importance Especially as Waterfowl Habitat", commonly known as the Ramsar Convention. This marked the beginning of our journey in wetland research and protection. Over the past thirty years, ecological science and wetland research in China have made significant progress; ecological civilization has become a national policy, enshrined in the Constitution, and public awareness of ecological protection continues to grow. Wetlands, as a vital and irreplaceable type of ecosystem, have gained wide recognition and protection from people all over the world, including the Chinese. In July 2024, the China Yellow (Bohai) Sea Migratory Bird Habitat (Phase II) was added to the World Heritage List, becoming one of the most important natural heritage sites globally, recognized and highlighted in the conservation of wetland ecosystems worldwide.

From another perspective, the Third Plenary Session of the 18th Central Committee of the Communist Party of China proposed the establishment of a national park system for the first time in 2013, making it a key task in the comprehensive deepening of reform. The construction of a characteristic national park system in China officially began. In October 2021, China officially established the first batch of five national parks: Sanjiangyuan, Giant Panda, Northeast Tiger and Leopard, Hainan Tropical Rainforest, and Wuyi Mountain, with a protected area of 230,000 square kilometers, covering nearly 30% of the country's key protected terrestrial wildlife species.

Thus, the construction of China's "natural protected area system based on national parks" has been significantly enhanced and improved, leading to proactive efforts at all levels, from national to local, in advancing the establishment of national parks. Among them, the planned "Yellow River Estuary National Park" is one of the most anticipated national parks!

According to my understanding, the Yellow River Estuary National Park will unify eight nature reserves, covering a total area of 2,812.90 square kilometers, which accounts for 79.96% of the park's total area. Given this scale, it's easy to understand that once established, the Yellow River Estuary National Park will more completely protect the estuarine wetland ecosystem of the Yellow River. Moreover, what is particularly valuable is that this wetland is one of the least populated areas along China's coast, meaning that various forms of wildlife, including birds, experience minimal human disturbance.

In recent years, institutions like the Institute of Zoology at the Chinese Academy of Sciences, along with other research organizations such as Beijing Normal University and the Chinese Academy of Environmental Sciences, have collaborated closely with local protected area authorities, wildlife management departments, and various levels of government to conduct comprehensive and in-depth research on the natural geography, ecological environment, and biodiversity of the Yellow River Estuary region.

Therefore, what I personally look forward to most is not just that excellent photographers from across the country come to the Yellow River Estuary to capture its stunning landscapes and the diversity of birds; more valuable and meaningful is that the public, especially young people, can enter the Yellow River Estuary National Park to appreciate the beauty of this natural heritage site and enrich their spiritual lives!

"When clouds and rain do disappear, all rainbows in these skies are clear. The sunset clouds in unison with lone fowls fly; The autumn water and vast skies are one in dye". As we admire this exquisite album, we should plan a journey for ourselves, to have the opportunity to come to Dongying, to visit the Yellow River Estuary National Park, and to experience the magnificence of wilderness nature and the elegance of flying birds.

Read this book, travel this path! This is the preface.

November 1, 2024

National Zoological Museum of China

CONTENTS

1

100个决定性瞬间

100 Decisive MOMENTS

自然奇观

Natural Wonders

014

2

100个决定性瞬间

100 Decisive MOMENTS

生命脉动

The Pulse of Life

066

3

100个决定性瞬间

100 Decisive MOMENTS

人文守望

Humanistic Concern

146

黄河口国家公园在哪里

Where is the Yellow River Estuary National Park?

East Longitude: 118°13'55" ~ 119°30'57"
North Latitude: 37°25'02"~ 38°17'53"

The park is located at the mouth of the Yellow River in Dongying City, Shandong Province, bordered to the north by the Bohai Sea, to the east by Laizhou Bay, and facing the Liaodong Peninsula across the sea.

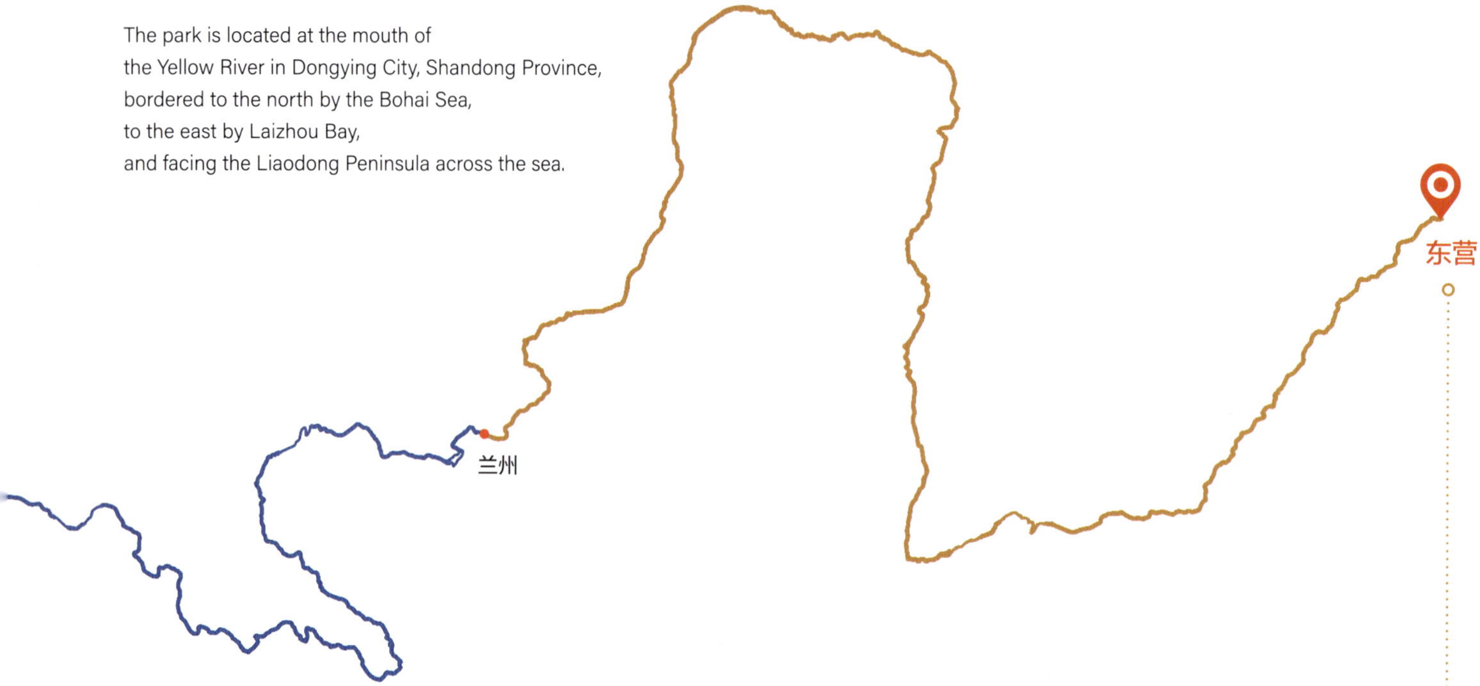

兰州

东营

东经 118° 13'55"~119° 30'57"
北纬 37° 25'02"~38° 17'53"

位于山东省东营市黄河入海口处，北临渤海，东靠莱州湾，与辽东半岛隔海相望。

河入海口

HOTO 张晓龙

PHOTO 张晓龙

黄河口：中国最美荒野

The Yellow River Estuary: China's Most Beautiful Wilderness

黄河口：中国最美荒野

人类创造了文明，文明却消灭了荒野。如今，野生动植物赖以生存的家园、作为“世界的保留地”的荒野，越来越稀缺。而黄河口，这块黄河入海、泥沙冲刷沉积而造就的新生土地，这片中国最年轻的湿地系统，已经成为中国最美的荒野。

黄河口国家公园总面积 3517.99 平方公里，是澳门特别行政区总面积的近 30 倍。其核心保护区面积 1842.44 平方公里，与韩国最大岛屿济州岛的面积相当。

黄河口这片广袤的荒野，拥有永久性河流、草本沼泽、灌木沼泽、内陆盐沼、潮间盐水沼泽、淤泥质海滩等丰富的河口湿地生态系统和海草床、浅海等典型的海洋生态系统，分布有大面积天然芦苇荡和柽柳林。黄河口国家公园记录有哺乳动物 24 种，鸟类 381 种，两栖爬行动物 16 种，鱼类 192 种，甲壳类动物 225 种，软体动物 168 种。

作为中国首个陆海统筹型国家公园，黄河口国家公园整合了黄河三角洲自然保护区、地质公园、森林公园、海洋特别保护区、水产种质资源保护区等 8 处自然保护地，陆域面积 1371 平方公里，海域面积近 2147 平方公里。

这块陆海交融的神奇土地，
处处彰显它的“荒野本色”。

The Yellow River Estuary: China's Most Beautiful Wilderness

Humanity created civilization, yet civilization has encroached upon the wilderness. Now, the habitats that wild flora and fauna rely on—the world's vital preserves—are becoming increasingly rare. But in the Yellow River Estuary, where the river meets the sea, the land newly formed by sediment deposits has emerged as China's youngest wetland system. This remarkable area stands as one of China's most stunning wild landscapes.

The Yellow River Estuary National Park covers an area of 3,518 square kilometers, approximately 30 times the total area of the Macao Special Administrative Region. Its core protected area spans 1,842.44 square kilometers, equivalent to the area of South Korea's largest island, Jeju Island.

This vast wilderness of the Yellow River Estuary features a rich estuarine wetland ecosystem, including permanent rivers, herbaceous wetlands, shrub wetlands, inland salt wetlands, intertidal saltwater wetlands, and muddy beaches, along with typical marine ecosystems such as seagrass beds and shallow seas. It is also home to extensive natural reed beds and tamarisk forests. The Yellow River Estuary National Park is recorded to host 24 species of mammals, 381 species of birds, 16 species of amphibians and reptiles, 192 species of fish, 225 species of crustaceans, and 168 species of mollusks.

As China's first national park integrating land and sea, the Yellow River Estuary National Park integrates eight natural protected areas, including the Yellow River Delta Nature Reserve, a geological park, a forest park, a marine special protection area, and a fishery germplasm resource protection area, with a terrestrial area of 1,371 square kilometers and a marine area of nearly 2,147 square kilometers.

This magical land where land and sea merge , showcases its "wilderness essence "everywhere.

荒野本色 之 新生湿地

黄河口是中国东部沿海地区人类活动最少的区域之一，是我国海陆变迁最活跃、面积增长速度最快的河口湿地生态系统，保留了完整的造陆过程，拥有中国暖温带最完整的河口湿地生态系统，是世界河口湿地生态系统的典型代表，是联合国环境署重点保护的全球13处湿地之一，“中国最美六大沼泽湿地”之一。

The Essence of Wilderness: The New Wetlands

The Yellow River Estuary is one of the least disturbed regions along China's eastern coastline. It is also the most dynamic estuarine wetland ecosystem in terms of land formation and expansion rate. Preserving an intact land-formation process, it hosts the most complete estuarine wetland ecosystem in China's warm temperate zone and serves as a global model for estuarine wetland ecosystems. Designated by the United Nations Environment Programme as one of the 13 priority-protected wetlands worldwide, it is also celebrated as one of "China's Six Most Beautiful Marsh Wetlands".

PHOTO 张晓龙

荒野本色 之 鸟的天堂

黄河口拥有我国暖温带保存最完整的湿地生态系统，为各种动植物尤其是鸟类的生长发育和繁衍生息创造了良好的条件，孕育了丰富的物种多样性，成为环西太平洋和东亚－澳大利西亚鸟类迁徙路线上重要的越冬地、中转站和繁殖地，在我国及全球生物多样性保护中占有极其重要的地位。

2024 年 7 月 26 日召开的第 46 届世界遗产大会上，山东东营黄河口候鸟栖息地成功入选《世界遗产名录》。

每年，数百万只鸟类在黄河口繁衍，这里因此享有“鸟类国际机场”的美誉。

The Essence of Wilderness: A Paradise for Birds

The Yellow River Estuary hosts China's most intact warm temperate wetland ecosystem, providing ideal conditions for the growth and thriving of diverse plants and animals, especially birds.It fosters rich species diversity and has become an important wintering ground, stopover, and breeding site along the migratory routes of birds in the West Pacific and East Asia-Australasia regions, playing an extremely important role in biodiversity conservation both in China and globally.

On July 26, 2024, at the 46th World Heritage Conference,The Yellow River Estuary Migratory Bird Habitat in Dongying, Shandong, has been successfully listed in *The World Heritage List*.

Every year, millions of birds stop, forage, and breed at the Yellow River Estuary, earning it the nickname "International Airport for Birds."

PHOTO 张晓龙

荒野本色 之 水生万物

黄河口是黄渤海区域海洋生物的重要种质资源库和生命起源地，也是其重要的产卵场、索饵场、越冬场和洄游通道。洄游性鱼类种类约占渤海区域的 74%，半滑舌鳎、蓝点马鲛等 40 多种鱼类在此产卵繁殖。

The Essence of Wilderness: Life Born from Water

The Yellow River Estuary is a crucial gene pool and cradle of marine life for the Yellow Sea and Bohai Sea regions. It serves as an important spawning ground, foraging area, wintering site, and migration passage. Anadromous fish species make up approximately 74% of the fish in the Bohai Sea region, with over 40 species, including the Chinese tongue sole and blue-spotted mackerel, spawning and reproducing here.

荒野本色 之 河海交汇

黄河口有黄河、渤海两大水文景观，黄河如同一条黄色巨龙，冲入碧蓝的渤海之中，在海面形成一道明显的黄蓝分界线，泾渭分明。

碧海、黄河在这里奇妙相遇，绘出一幅“蓝黄交融”的神奇画卷，也是世界独特的自然奇观。

The Essence of Wilderness: Where Rivers Meet the Sea

The Yellow River Estuary presents two major hydrological landscapes: the Yellow River and the Bohai Sea.The Yellow River, like a massive yellow dragon, flows into the azure Bohai Sea, creating a distinct yellow-blue boundary on the water's surface, clearly delineating the two.

Here, the azure sea and the yellow river meet in a marvelous way, creating an enchanting scene of "blue meeting yellow" which is a unique natural wonder of the world.

PHOTO **赵英丽**

这就是黄河口，中国最美的荒野。

This is The Yellow River Estuary, the most beautiful wilderness in China.

PHOTO 张晓龙

在生态学家奥尔多 · 利奥波德倡导的“大地伦理”中，作为生态系统的自然是一个呈现着美丽、完整与稳定的生命共同体。荒野自然有一种完整性，它独立运行着，不在我们的掌握之中，而这，也正是它的价值所在。

面对这片荒野，我们所能做的是，
远离它，让荒野永远归荒野；
走向它，去寻找和聆听它以自然
的形式表达自己，
在野性的自然和广袤的荒野中，
发现并呈现不依赖于人类而存在
的原初之美，用“决定性瞬间”
定格自然的每一次生命悸动。

In Aldo Leopold's advocacy of the "Land Ethic", nature, as an ecosystem, is viewed as a community of life that embodies beauty, integrity, and stability. Wilderness possesses an inherent wholeness, operating independently, beyond human control—and therein lies its true value.

In the face of this wilderness, what we can do is to either distance ourselves from it, allowing it to remain wild forever, or to approach it, seeking to listen to its expression of self in its natural form. In the wildness of nature and the vastness of the wilderness, we can discover and present a primal beauty that exists independent of humanity, capturing the "Decisive Moment" of every natural life.

1 自然奇观

黄河从中游黄土高原地区携带来的大量泥沙在黄河口淤积造陆形成了黄河三角洲。

这片年轻的荒野，在河与海的共同孕育下，形成了独特的地形地貌，造就了以“奇、特、旷、野、新”为主要美学特征的自然景观。

Natural Wonders

The Yellow River carries a significant amount of sediment from the Loess Plateau as it flows through the middle reaches, which accumulates at the Estuary to form the Yellow River Delta.

This young wilderness, nurtured by both river and sea, has created a unique topography and geomorphology, resulting in a natural landscape characterized by the aesthetic features of "wonder, uniqueness, vastness, wilderness, and novelty."

100个
决定性瞬间

100 Decisive Moments

黄河口国家公园

The Yellow River Estuary National Park

MOMENT 1

黎明前的黄河口

河流在滩涂上静静地流淌，鸟儿安然地享受着荒野馈赠的万籁俱寂。这一画面让人不觉联想到聂鲁达著名的诗句：“我喜欢你是寂静的。”

The Yellow River Estuary Before Dawn

The river quietly flows across the tidal flats, while the birds peacefully enjoy the tranquility gifted by the wilderness. This scene evokes the renowned line from Pablo Neruda's poem: "I like you when you are silent."

PHOTO 赵文昌

MOMENT 2

苏醒的土地

太阳从地平线升起，将一群鸬鹚从睡梦中叫醒。崭新的一天，就要开始了。黄河水带来的泥沙造就了新的滩涂，不断向大海推进。完整的生态系统使黄河入海口能够按照自然演替规律进行能量流动和物质循环。

Awakening Land

The sun rises above the horizon, waking a flock of cormorants from their slumber. A brand new day is about to begin. The sediment carried by the Yellow River gives rise to new tidal flats, steadily advancing toward the sea. A complete ecological system allows the Yellow River Estuary to facilitate energy flow and material cycling according to the natural succession processes.

PHOTO 张树岩

MOMENT 3

黄河口日出

在黄河入海的地方观东方破晓，分外奇伟壮观。万道亮丽的霞光，将夜的帷幕拉开。

太阳钻出云层，冉冉升起，金色的光柱射向大地，滋养沃土息壤。

The Sunrise over the Yellow River Estuary

Witnessing the dawn from where the Yellow River meets the sea is truly magnificent. A thousand beams of radiant light draw back the curtain of night.

The sun emerges from the clouds, rising slowly, casting golden rays upon the earth, nourishing the fertile soil.

黄河口日出时间

Sunrise Times at the Yellow River Estuary

3月20或21日 **06:06**

Spring Equinox

6月21或22日 **04:44**

Summer Solstice

9月22或23日 **05:53**

Autumn Equinox

12月21或22日 **07:16**

Winter Solstice

PHOTO 钱茜

MOMENT 4

在河海交汇时相遇

一只红嘴巨燕鸥在河海交汇处飞过，此时的海面上，一艘游轮也正好在黄蓝融合线上驶过。

山东省旅游局十大文化旅游目的地品牌，“黄河入海”位列其中。黄河，中华民族的母亲河，华夏文明的摇篮，流经我国 9 省（自治区），在山东东营入海，形成了壮美的自然景观。

在无人机捕捉的画面中，河海交汇如一条黄蓝飘带，把浑浊的黄河水与碧蓝的海水劈为两半，河黄海蓝。

河海交汇每天都在黄河入海口上演，但受黄河水量等客观因素制约，每年能从黄河乘船出海观赏河海交汇的几率比登泰山看日出的几率还要低。

Encountering at the River-Sea Confluence

A Caspian Tern flew over the confluence of the river and sea, while a cruise ship passed exactly along the line where yellow and blue merge on the water's surface.

“神舟十四号”航天员陈冬在太空拍摄的黄河入海照片

“The Yellow River Entering the Sea” photo taken by astronaut Chen Dong of "Shenzhou-14" in space.

PHOTO 胡友文

Meandering Flows on the Tidal Flats

Various streams meander across the tidal flats of the Yellow River Estuary, creating a unique topography and ecosystem. They twist and wind, exuding a tranquil beauty as they flow gently toward the sea.

PHOTO 赵文昌

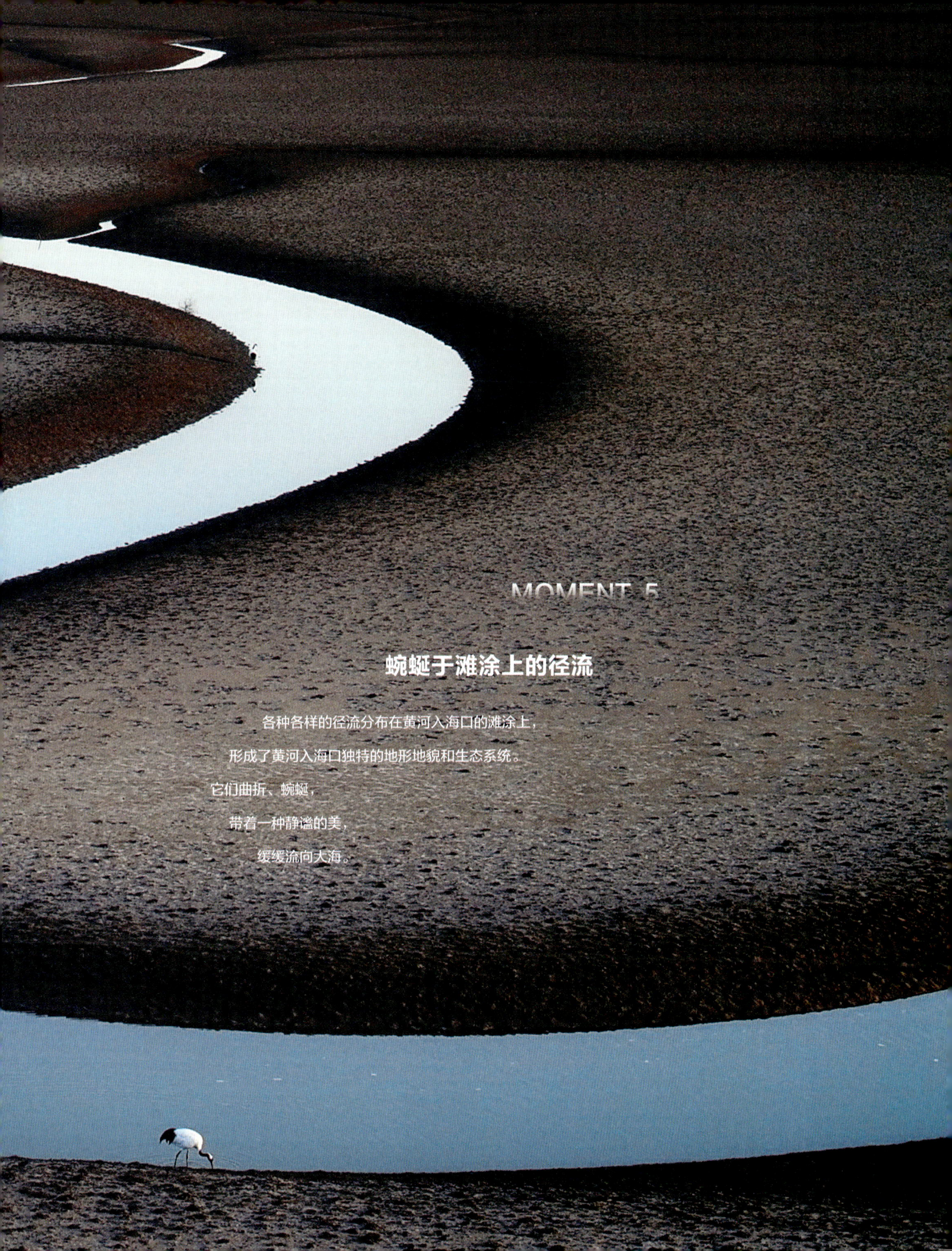
MOMENT 5
蜿蜒于滩涂上的径流
各种各样的径流分布在黄河入海口的滩涂上，
形成了黄河入海口独特的地形地貌和生态系统。
它们曲折、蜿蜒，
带着一种静谧的美，
缓缓流向大海。

MOMENT 6

鹤翔黄河口

独特的地形地貌和自然条件造就了以“奇、特、旷、野、新”为主要特点的黄河口景观。

此刻，你可以想象自己就是一只翱翔于天际的丹顶鹤，在你的目光之下，大自然神奇的造物一览无遗。你怎能不为之倾心动容？

Crane Soaring Over the Yellow River Estuary

The unique topography and natural conditions of the Yellow River Estuary create a landscape characterized by its "strangeness, uniqueness, vastness, wilderness, and novelty."

At this moment, you can imagine yourself as a soaring Red-crowned Crane, taking in the breathtaking wonders of nature below. How could you not be captivated by its beauty?

PHOTO **赵文昌**

MOMENT 7

发现新大陆

黄河口是一块仍在不断生长的土地。黄河作为世界上输沙量最大的河流，多年均有 10 亿多吨泥沙在入海口沉积，多年均造陆 20 多平方公里、河道向海中延伸超过 2 公里，使黄河口成为世界上面积增长最快的土地。

Discovering a New Land

The Yellow River Estuary is a land that continues to grow. As the river with the largest sediment discharge in the world, the Yellow River deposits over 1 billion tons of sediment at its mouth each year. This accumulation leads to the creation of more than 20 square kilometers of land annually, with the river channel extending over 2 kilometers into the sea, making the Yellow River Estuary the fastest-growing landmass in the world.

PHOTO 李刚

MOMENT 8

大地色彩

绚烂的红，

温暖的橙，

清新的蓝，

生机勃勃的绿，

璀璨闪耀的金……

黄河口湿地仿佛打翻了大自然的调色板，

绘出一幅色彩斑斓的印象派油画。

PHOTO 黄高潮

Colors of the Earth
Vibrant reds,
Warm oranges,
Refreshing blues,
Lively greens,
Brilliant shining golds...
The Yellow River Estuary wetland seems to have overturned nature's palette,
Creating a dazzling impressionist masterpiece filled with a riot of colors.

MOMENT 9

大地肌理

盐碱地、湿地、海滩，共同构成了黄河口的大地肌理。

黄河三角洲地区的盐碱地面积高达600多万亩，其形成主要因为黄河水中带有大量泥沙，其中包括盐碱成分，这些成分在沉积过程中逐渐积累。同时，海水的浸灌和随后的蒸发也会导致盐碱成分在地表富集。

Textures of the Earth

Saline-alkaline land, wetlands, and beaches collectively form the unique texture of the Yellow River Estuary. The saline-alkaline land in the Yellow River Delta spans over 6 million mu, primarily formed by the large amount of sediment, inclusive of saline and alkaline components, carried by the Yellow River. These components gradually accumulate during the sedimentation process.

PHOTO 赵文昌

MOMENT 10

群鸟栖居“地球之肾”

这个由摄影师使用无人机抓取的惊艳绝伦的瞬间，完美呈现了鸟群在黄河口湿地中栖居的状态。

湿地与森林、海洋并称为“全球三大生态系统”，孕育和丰富了全球的生物多样性，被称为“地球之肾”。

2013 年，黄河三角洲被列入“国际重要湿地名录”。

Birds Nesting in the "Kidneys of the Earth"

This breathtaking moment, captured by the photographer using a drone, perfectly presents the state of bird flocks residing in the Yellow River Estuary wetlands.

Wetlands, alongside forests and oceans, are known as the "three major ecosystems of the planet," nurturing and enriching global biodiversity, and are often referred to as the "kidneys of the Earth."

In 2013, the Yellow River Delta was included in the "List of Internationally Important Wetlands."

PHOTO **赵文昌**

MOMENT 11 诗意家园

一群灰鹤在黄河口湿地的芦苇荡上空飞翔。作为人类，你是不是会羡慕这些栖居于“中国最美湿地”中的自由的生命?

在湿地存在形态上，黄河三角洲湿地以常年积水湿地为主，占总面积的 63%，且滩涂湿地在其中占优势地位。季节性积水湿地占湿地总面积的 37%。

A Poetic Homeland

A flock of Sommon Cranes soars over the reed wetlands of the Yellow River Estuary. As a human, don't you envy these free beings that inhabit "China's Most Beautiful Wetland"?

In terms of the forms of wetlands, the Yellow River Delta mainly comprises permanently flooded wetlands, which make up 63% of the total area, with tidal wetlands having a dominant position within that. Seasonal flooded wetlands make up the remaining 37% of the wetland area.

PHOTO 禹明善

MOMENT 12

一棵树

这是一棵“潮汐树”。潮汐树，黄河三角洲一种典型的沉积地貌。由潮汐反复冲刷而形成的一条条潮沟，犹如生长在滩涂上的参天大树，其“树干”朝向大海，“枝杈”朝向陆地。

A Tree

This is a "Tidal Tree." The Tidal Tree is a typical sedimentary landform in the Yellow River Delta. Formed by the repeated scouring of tides, the tidal channels resemble towering trees growing on the mudflats, with their "trunks" facing the sea and their "branches" extending toward the land.

PHOTO **秦金武**

"潮汐树"形成示意图

Diagram of "Tidal Tree" Formation

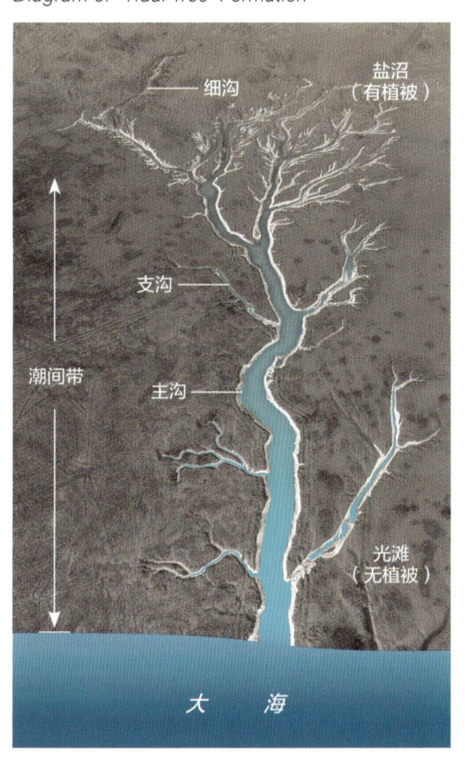

MOMENT 13

一片羽

严寒冬日，大自然在黄河口的滩涂上，以冰雪为底色，在"潮汐树"上尽情挥毫泼墨，勾勒出一片水墨丹青的羽毛。

A Feather

On a frigid winter day, nature paints on the mudflats of the Yellow River Estuary with ice and snow as its canvas, creating a water-and-ink masterpiece of feathers upon the "Tidal Tree."

PHOTO **赵文昌**

MOMENT 14

粉紫色的春天

每年 4 月起，罗布麻的粉紫色小花会开满盐碱滩涂及湿地河道两侧。

罗布麻是一种生长于盐碱土地上的多年生宿根草本植物，在黄河口地区有广泛分布，因其叶片具有茶的功效，当地人称之为“茶棵子”，对治疗高血压、心力衰竭等疾病有较好的疗效。

Powdery Purple Spring

Every April, the powdery purple flowers of the ramie plant bloom abundantly along the saline-alkaline mudflats and the banks of wetland rivers.

Ramie is a perennial herbaceous plant that grows in saline-alkaline soils and is widely distributed in the Yellow River Estuary region. Its leaves resemble those of tea, leading locals to affectionately refer to it as "tea tree." This plant is known for its efficacy in treating conditions such as hypertension and heart failure.

PHOTO 孙劲松

MOMENT 15

绿波盛夏

夏日的黄河口绿波荡漾，水草依依随风摇曳，欢快的鸟儿在天际自由地飞翔。

河水流经湿地时，其中所含的营养成分被湿地植被吸收，净化了下游水源。

Green Waves of Midsummer

In the summer at the Yellow River Estuary, green waves ripple gently as aquatic plants sway in the breeze, while cheerful birds soar freely across the sky.

As the river flows through the wetlands, the nutrients it carries are absorbed by the wetland vegetation, purifying the water for downstream areas.

PHOTO 丁洪安

MOMENT 16 秋夏之交 色彩斑斓

秋夏交替之时，是黄河口色彩最为丰富的时节。红黄蓝绿，五彩斑斓，鸟儿都仿佛被眼前的景色吸引，驻足欣赏。

每年 8 月至 11 月，黄河口进入“最美”的时间。

Colorful Transition of Late Summer to Autumn

As summer gives way to autumn, the Yellow River Estuary bursts into a vibrant palette of colors. Shades of red, yellow, blue, and green create a kaleidoscope of hues, captivating the birds that pause to admire the stunning scenery.

From August to November each year, this period is heralded as the "most beautiful" time at the Yellow River Estuary.

PHOTO 丁洪安

MOMENT 17 秋天的“红地毯”

每年入秋伊始，绵延不绝的碱蓬草把广袤的滩涂染成红色，成为秋季黄河口最为独特的风景。人们把这一景象形象地称为“红地毯”。

碱蓬草是一种生长于高盐碱地的植物，以其独特的红色叶片和适应极端环境的特性而闻名，是黄河三角洲湿地上的标志性植被之一。

Autumn's "Red Carpet"

Every year at the beginning of autumn, the sprawling Suaeda covers the vast mudflats in a blanket of red, becoming the most unique scenery at the Yellow River Estuary in autumn. People vividly refer to this phenomenon as the "red carpet."

初春，刚刚发芽的碱蓬草

Early spring, the newly sprouted Suaeda.

PHOTO 杨霞

PHOTO 张晓龙

MOMENT 18

芦荻飞雪

芦荻像大地的诗行，秋日的童话，构成了黄河入海口深秋的文艺美学。

浩瀚的芦苇荡是黄河三角洲极具特色的景观。由于芦苇的叶、叶鞘、茎、根状茎和不定根都具有通气组织，所以它在净化污水中起到重要的作用。

Snow-like Reeds Flying in the Air

Reeds are like the verses of the earth, a fairy tale of autumn, creating an artistic aesthetic in the late autumn at the Yellow River Estuary.

The vast expanse of reeds is a distinctive landscape of the Yellow River Delta. The leaves, leaf sheaths, stems, rhizomes, and adventitious roots of reeds all have aerenchyma, which plays a crucial role in purifying wastewater.

黄河口四季中的芦荻

The Reeds of the Yellow River Estuary Through the Seasons

PHOTO **杨霞**

春天芦笋破土，生机盎然； In spring, the reed shoots break through the soil, bursting with vitality;

夏天芦荡茫茫，碧波翻滚； In summer, the reed beds stretch endlessly, with rolling emerald waves;

秋季芦荻飘絮，纷纷扬扬； In autumn, the fluffy reeds flutter, dancing in the breeze;

冬季芦苇盖雪，金白辉映。 In winter, the reeds are blanketed in snow, shining in golden-white hues.

PHOTO 张目

MOMENT 19

飞越芦苇荡

一只东方白鹳从芦苇荡上飞过。一望无际的芦苇荡为鸟类提供了天然的保护屏障，成为它们安全可靠的栖息地。

Soaring Over the Reed Beds

An Oriental White Stork glides over the vast expanse of the reed wetlands. The endless reeds create a natural protective barrier for the birds, making it a safe and reliable habitat for them.

PHOTO **赵文昌**

MOMENT 20

金色大地上的生命之树

深秋时节，群鸟降落在海滩“潮汐树”的枝杈上，如同树上挂满了果实。被摄影师捕捉到的这个精彩瞬间，为寂寥的秋日注入了生机与活力。

The Tree of Life on Golden Land

In late autumn, flocks of birds perch on the branches of the "Tidal Tree" along the beach, resembling fruit hanging from its limbs. This stunning moment, captured by the photographer, brings vitality and energy to the solitary autumn day.

PHOTO 秦金武

MOMENT 21

雪中鹤

数以万计的鸟会在冬季到来前成群结队飞抵黄河口，将这里作为越冬地和迁徙的中转站。蒹葭苍苍，雁鸣鹤舞，将黄河口的冬日渲染得如诗如画。

Cranes in the Snow

As winter approaches, thousands of birds flock to the Yellow River Estuary, using it as a wintering ground and a stopover on their migration route. Amidst the lush reeds and the calls of geese and cranes, the winter days at the estuary are painted in a poetic and picturesque light.

PHOTO **赵文昌**

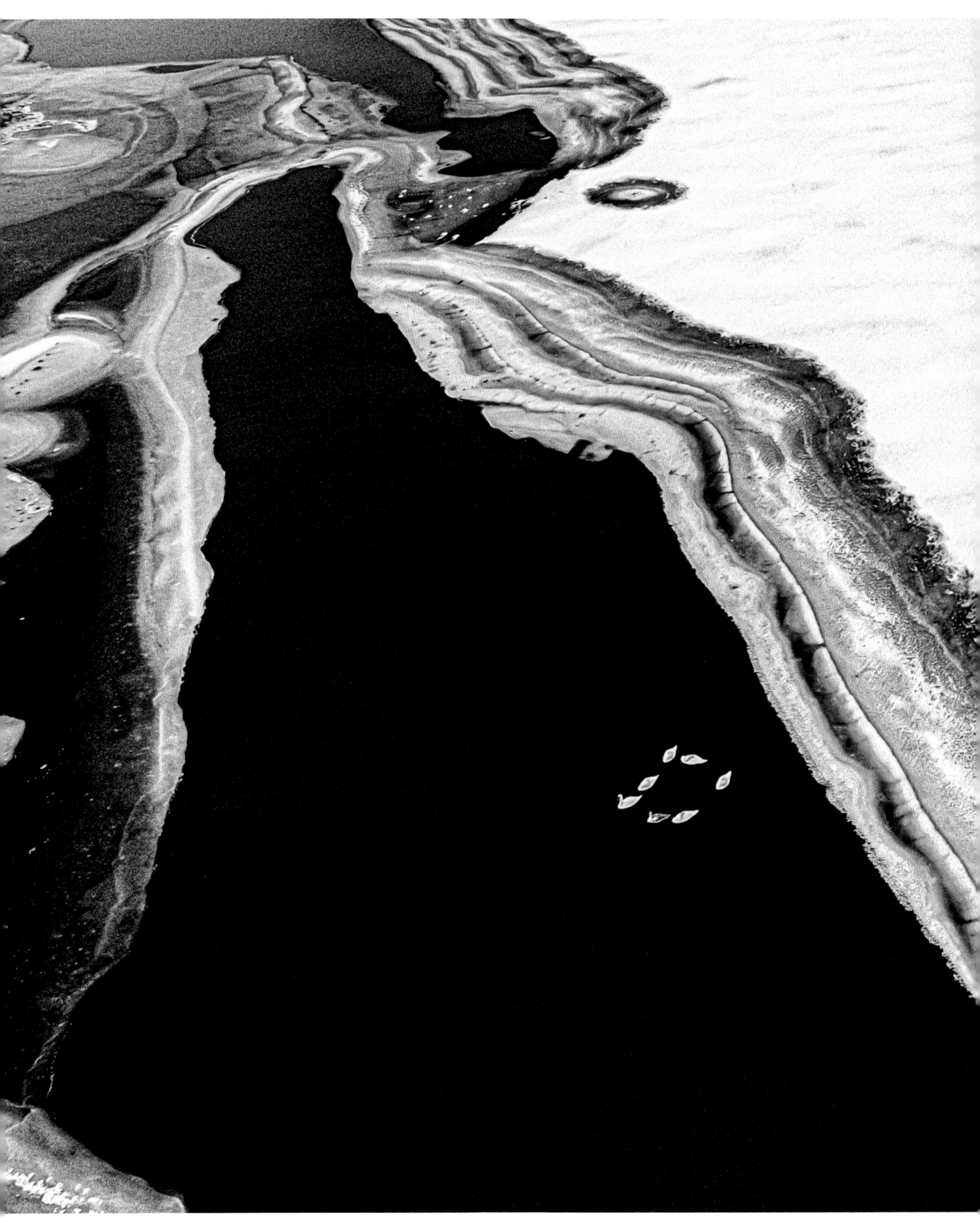

MOMENT 22

冬日恋歌

冬季的黄河三角洲冰雪一色，成为天鹅理想的越冬之地。黄河三角洲属于温带季风气候，四季分明。 冬季大部分河流因为流速缓慢，都会结冰，让大地呈现出纯粹又纯净的美感。

Winter Love Song

In winter, the Yellow River Delta is blanketed in ice and snow, becoming an ideal wintering ground for swans. The Yellow River Delta enjoys a temperate monsoon climate with distinct seasons. During winter, most rivers slow down and freeze, allowing the land to display a pure and pristine beauty.

PHOTO 张晓龙

MOMENT 23

“3”生万物

一群东方白鹳停歇在黄河河滩上，神奇般地摆出了阿拉伯数字“3”的造型。

荒野中出现的这一幕，仿佛是对“三生万物”的哲学解读——万物的产生和发展基于原始的、混沌的状态，经过一系列的变化和演化，最终形成了丰富多彩的物质世界。让自然保持其荒野价值，就是人类在守护生命之源。

"3" Gives Rise to All Things

A flock of Oriental White Storks rests on the banks of the Yellow River, miraculously forming the shape of the Arabic numeral "3." This scene in the wilderness seems to echo the philosophical interpretation of "Three Gives Rise to All Things" — that the emergence and development of all things are rooted in an original, chaotic state. Through a series of changes and evolutions, this leads to the vibrant material world we know. Preserving the wild value of nature is humanity's way of safeguarding the source of life.

PHOTO 孙守山

MOMENT 24

在黄河之上

它从河水与河床明暗分明的两种黄中冲出来，翱翔在黄河之上，如此自由，如此野逸。这种时刻，很容易被我们投射到自身，启迪着我们思考生命的意义。这就是荒野存在的价值，它是人类精神的摇篮。

Above the Yellow River

It emerges from the contrasting yellows of the river water and riverbed, soaring above the Yellow River, so free and wild. Moments like this easily resonate within us, inspiring reflections on the meaning of life. This is the value of the wilderness; it is the cradle of the human spirit.

PHOTO **焦搏**

MOMENT 25

大河尽头

九曲黄河，奔腾万里，从东营注入渤海。大河流淌 5464 公里，在黄河口进入了它漫长奔波的终点。它卸下了一身疲惫，放缓了脚步，仿佛要享受这段最后的旅程。

今黄河东营段，上起滨州界，自西南向东北横贯东营市全境，在垦利区东北部注入渤海，全长 138 公里。

End of the Great River

The meandering Yellow River rushes for thousands of miles, finally pouring into the Bohai Sea at Dongying. Flowing a total of 5,464 kilometers, it reaches the end of its long journey at the estuary, shedding its fatigue and slowing down, as if to savor this final leg of the voyage.

The Dongying section of the Yellow River extends 138 kilometers, starting at the Binzhou boundary and traversing the entire city from southwest to northeast, finally entering the Bohai Sea in the northeastern part of Kenli District.

PHOTO 张晓龙

MOMENT 26

暮色中的黄河故道

黄河故道依旧保持着未被开垦的原始风貌，独守着一份不受尘世纷扰的宁静。

由于黄河特有的河情，历史上洪水灾害频繁，以“善淤、善决、善徙”而著称，向有“三年两决口，百年一改道”之说。

黄河入海的流路按照淤积→延伸→抬高→摆动→改道的规律不断演变，使黄河三角洲的面积不断扩大，海岸线不断向海推进，历经 160 余年，逐渐淤积成近代黄河三角洲。

The Yellow River Old Channel at Dusk

The old channel of the Yellow River still preserves its untouched, primitive appearance, safeguarding a tranquility that remains unperturbed by the chaos of the world.

Due to the unique characteristics of the Yellow River, historical flooding disasters have been frequent, earning it the reputation of being "prone to sedimentation, breach, and diversion," with the saying, "once every three years there is a breach, and every century a change in the course."

The flow path of the Yellow River into the sea continually evolves according to the patterns of sedimentation → extension → elevation → fluctuation → diversion, leading to the constant expansion of the Yellow River Delta and the shoreline advancing further into the sea. Over the course of more than 160 years, this process has gradually formed the modern Yellow River Delta.

PHOTO 丁洪安

PHOTO 胡友文

MOMENT 27

流金溢彩的黄河入海口

黄昏时分，一只丹顶鹤在入海口的滩涂上悠闲踱步。

夕阳下，黄河滩涂的河床充满了金属的质感。这一刻，自然呈现出母亲般的温暖与柔情。被人类拟人化的“自然母亲”在这里找到了具体的形象。

The Golden and Colorful Yellow River Estuary

At dusk, a Red-crowned Crane strolls leisurely across the mudflats of the estuary.

In the setting sun, the riverbed of the Yellow River's mudflats radiates a metallic sheen. In this moment, nature reveals a warmth and tenderness reminiscent of a mother. The personified "Mother Nature" finds a concrete form here, embodying nurturing qualities that resonate deeply with humanity.

PHOTO 赵文昌

MOMENT 28

落日熔金

夕阳的余晖倾洒而下，金色的光芒如同细密的纱幔，轻柔地覆在河滩上，鸟群似乎不舍归巢，仿佛在等待一场落日音乐会的开始。

The Setting Sun Melts Gold

The afterglow of the setting sun tumbles down, enwrapping the riverbank in a golden light that resembles a delicate veil. The flock of birds seems reluctant to return home, as if awaiting the beginning of a sunset concert.

MOMENT 29

晚安，黄河口！

夜深了，倦鸟已然入睡，黄河口安静下来。“湿地家园”这个词汇中所蕴含的“家”的意义，被这个画面完美地传递出来。

Goodnight, Yellow River Estuary!

As night falls, the weary birds have settled into sleep, and the Yellow River Estuary grows quiet. The essence of "home" embedded in the term "wetland habitat" is perfectly conveyed through this scene.

PHOTO 张晓龙

The Pulse of Life

The Yellow River Estuary lies at a crucial midway point and passageway along the Northeast Asia Inland and Western Pacific migratory bird routes. It is a vital stopover, breeding ground, and wintering site for migratory birds. This unique geographic position and rich ecological environment endow the delta with an abundance of avian resources, welcoming millions of birds that halt here each year during their migration. Birds are the true inhabitants of the Yellow River Estuary. This is their world, where life pulses on with every wingbeat.

生命脉动

黄河口地处东亚－澳大利西亚和环西太平洋鸟类迁徙路线的中间环节和咽喉要道，是候鸟迁徙、繁殖、越冬的必经之地。独特的地理位置、优越的生态环境赋予了黄河口丰富的鸟类资源，每年在这里迁徙停歇的鸟类达数百万只。鸟是黄河口的主人。这里是鸟的世界，生命在飞翔中脉动不息。

100个决定性瞬间

100 Decisive Moments

黄河口国家公园

The Yellow River Estuary National Park

MOMENT 30

天堂晨曦

PHOTO 张东河

博尔赫斯想象中的天堂，是图书馆的模样。如果真的有天堂，它或许就是这幅画面吧！假如天堂有日出，它应该就是这个样子吧！这一刻，朝阳、湿地、鸟群、晨雾共同构成了一把打开我们心灵的自然钥匙。

由于黄河口湿地生态资源得天独厚，这里已经成为近 100 种国家重点保护鸟类的重要繁殖地、越冬地和迁徙停歇地。对鸟类而言，这里就是它们生命的天堂。

Paradise at Dawn

Jorge Luis Borges imagined paradise as a library. But if paradise truly exists, it might look like this scene. If there were a sunrise in paradise, it would appear just like this. In this moment, the sunrise, wetlands, flocks of birds, and morning mist together form a natural key that unlocks our souls.

Due to its rich ecological resources, The Yellow River Estuary wetland has become a vital breeding ground, wintering site, and migratory stopover for nearly 100 species of nationally protected birds. For the birds, this place is indeed their paradise.

MOMENT 31

万鸟竞飞

上万只鸻鹬类水鸟在黄河口的水面上飞舞。鸻鹬类是湿地鸟类中最多的类群。

黄河三角洲广阔的滩涂和浅海湿地吸引了数以百万计的鸟类在此停留栖息，被鸟类专家誉为“鸟类的国际机场”。

Thousands of birds compete to fly

Tens of thousands of shorebirds flutter above the waters of the Yellow River Estuary, a site dominated by this largest group of wetland birds.

The vast mudflats and shallow coastal wetlands of the Yellow River Estuary attract millions of birds to pause and rest here, earning it the title "International Airport for Birds" from avian experts.

PHOTO 刘月良

PHOTO 黄高潮

MOMENT 32

鸟图腾

两只东方白鹳在阳光的映衬下，幻化成一幅鸟的图腾。黄河口是全球最重要的东方白鹳繁殖地，被誉为“中国东方白鹳之乡”。

2003 年 4 月，在黄河口管理站垦东五附近发现了第一窝东方白鹳繁殖巢。自此之后，东方白鹳繁殖种群逐年增长。截至 2024 年，黄河口已累计繁殖东方白鹳雏鸟 3724 只，为保护这一物种做出了突出贡献。

Bird Totem

In the sunlight, two Oriental White Storks as if transformed into a totem of birds. The Yellow River Estuary, the world's most crucial breeding ground for Oriental White Storks, is celebrated as the "Homeland of the Oriental White Stork" in China.

In April 2003, the first breeding nest of the Oriental White Stork was discovered near the Kengdong 5 area at the Yellow River Estuary management station. Since then, the breeding population of Oriental White Stork has steadily grown each year. By 2024, The Yellow River Estuary has successfully nurtured a cumulative total of 3,724 Oriental White Storks, making a remarkable contribution to the preservation of this species.

东方白鹳

一种大型涉禽，体长为 110–128 厘米，体重 3.9–4.5 千克，翼展宽大约 2.22 米。其黑色喙粗大，除飞羽黑色外全身白色。属国家一级保护动物，被国际自然保护联盟定为濒危物种。

Oriental White Stork

The Oriental White Stork is a large wading bird, measuring 110-128 cm in length and weighing between 3.9-4.5 kg, with a wingspan of about 2.22 m. It has a large black beak, and its body is predominantly white, except for the black flight feathers. This species is classified as a national first-class protected animal and is listed as endangered by the International Union for Conservation of Nature (IUC).

PHOTO **付建智**

A Thousand Storks in Sight

This moment was captured by the photographer: thousands of Oriental White Storks foraging at the same time in the wetlands of the Yellow River Delta, creating a breathtaking spectacle.

MOMENT 33

一目千鹳

这是摄影师无意间捕捉到的一个瞬间：数千只东方白鹳同时在黄河口的湿地上觅食，甚为壮观。

观东方白鹳指南

在黄河口观东方白鹳，除冬季湿地大面积结冰外，一年四季皆可见。在春、秋迁徙季，可见数量众多的东方白鹳；繁殖季，东方白鹳繁衍生息的实景则难得一见。

Guide to Viewing Oriental White Storks

At the Yellow River Estuary, Oriental White Storks can be observed throughout the year, except in winter when the wetlands freeze extensively. During the spring and autumn migration seasons, large numbers of Oriental White Storks can be seen. However, witnessing their breeding activities during the nesting season is a rare experience.

PHOTO 赵文昌

MOMENT 34

鹤生于此

这是 2023 年黄河口国家公园的科研人员在进行野外监测的时候拍摄到的丹顶鹤野外自然繁殖巢。

丹顶鹤野外自然繁殖于 2019 年在黄河三角洲首次被记录，这里也成为丹顶鹤在我国越冬的最北境、自然繁殖的最南境。

Cranes Born Here

This is a photograph taken by researchers during field monitoring in 2023, showing a wild breeding nest of the Red -Crowned Crane.

The first record of Red -Crowned Cranes breeding in the wild in the Yellow River Estuary was made in 2019, making this area the northernmost wintering ground and the southernmost natural breeding ground for Red -Crowned Cranes in China.

丹顶鹤

又名仙鹤，是鹤类中的一种大型涉禽，体长 120-160 厘米，常被人冠以“湿地之神”的美称。作为国家一级保护动物，在全球的野外种群数量只有 2800-3000 只。仙鹤是中国文化中的一个重要的符号，在古代，它是仅次于凤凰的“一鸟之下，万鸟之上”的“一品鸟”，明清一品官吏的官服编织的图案就是“仙鹤”。还由于寿命长达 50-60 年，人们常把它作为长寿的象征。仙鹤独立，翘首远望，姿态优美，色彩不艳不娇，是高贵典雅的象征。

Red -Crowned Crane

Also known as the "immortal crane," the Red -Crowned Crane is a large wading bird, measuring 120-160 cm in length, often referred to as the "god of wetlands." As a national first-class protected animal, its global wild population is only 2,800-3,000 individuals.

The crane holds significant cultural symbolism in China; in ancient times, it was regarded as the "bird of excellence," ranking just below the phoenix. The patterns woven into the official robes of first-class officials during the Ming and Qing dynasties featured cranes. With a lifespan of 50-60 years, the crane is often seen as a symbol of longevity. Standing gracefully and gazing into the distance, it embodies elegance and nobility, with its understated colors symbolizing refinement.

PHOTO 吴立新

MOMENT 35

鹤鸣长歌于九皋

两对丹顶鹤在沼泽中仰天长歌，这一画面就是《诗经》中“鹤鸣于九皋，声闻于天”的生动展现。

丹顶鹤需要洁净而开阔的湿地环境作为栖息地，是对湿地环境变化最为敏感的指示生物。

Cranes Sing in the Marshlands

Two pairs of Red-Crowned Cranes sing joyfully towards the sky in the marsh, vividly illustrating the lines from the *Book of Songs*: Cranes call in the nine marshes, their voices heard in the heavens.

Red-Crowned Cranes require clean and open wetland environments as their habitat, making them highly sensitive indicators of changes in wetland conditions.

PHOTO 孙劲松

MOMENT 36

天鹅湖

湿地湖泊里嬉戏的一群天鹅。时而在水中游弋，时而在空中飞舞。

黄河口良好的湿地生态，为天鹅提供了大量的水草、种子、昆虫等充足的食物，湿地芦苇浩荡、河流交叉纵横，是天鹅理想的越冬之地。

Swan Lake

A group of swans plays merrily in the wetland lake, gliding gracefully through the water and soaring through the sky.

The abundant wetland ecology of the Yellow River Estuary provides swans with ample food, including aquatic plants, wild grasses, seeds, and insects. The vast reed beds and intricate network of rivers make this area an ideal wintering habitat for swans.

亚洲及黄河三角洲的天鹅

世界上共 7 种天鹅，亚洲有 3 种，分别为：大天鹅、小天鹅、疣鼻天鹅。在中国，所有天鹅为国家二级保护动物。3 种天鹅在一处地方同时出现并不常见，黄河口是能同时见到 3 种天鹅的少数区域之一。

Swans of Asia and The Yellow River Estuary

There are seven species of swans in the world, three of which are found in Asia: the Whooper Swan, the Tundra Swan, and the Mute Swan. In China, all swan species are classified as second-class protected animals. It is rare for all three species to be seen in one location, making the Yellow River Estuary one of the few areas where all three can be observed simultaneously.

大天鹅 Whooper Swan

小天鹅 Tundra Swan

疣鼻天鹅 Mute Swan

PHOTO 刘月良

PHOTO 丁洪安

MOMENT 37

迁徙之路

一群丹顶鹤在黄河口湿地上空飞翔。丹顶鹤在每年春秋迁徙季节路过黄河口，自 2019 年以来，部分丹顶鹤选择黄河口繁殖。

在世界鸟类 9 条迁徙路线中，黄河口横跨东亚 – 澳大利西亚和环西太平洋 2 条迁徙路线；在中国 4 条迁徙路线中，处于东部候鸟迁徙路线的咽喉要道，是候鸟迁徙的必经之路。

Migration Routes

A flock of Red-Crowned Cranes soars above the wetlands of the Yellow River Estuary. During their annual spring and autumn migration seasons, Red-Crowned Cranes pass through this area. Since 2019, some have chosen to breed here.

The Yellow River Estuary spans two of the nine major migratory routes for birds globally: the East Asia-Australasia route and the Pacific Flyway. It is a vital corridor in China's eastern migratory routes, making it an essential stopover for migratory birds.

PHOTO 丁海

MOMENT 38

百“家”争鸣

摄影师偶遇的一个位于黄河故道的苍鹭巢群。苍鹭喜欢在湿地附近的岩壁或树上集群营巢。但几百个苍鹭“家庭”聚集在一起振翅争鸣的场景实属罕见，蔚为壮观。

A Symphony of Herons

The photographer encountered by chance a colony of Grey Heron nests situated along the Yellow River's ancient riverbed. Grey Herons typically prefer to nest in groups on rock cliffs or trees near wetlands. However, witnessing hundreds of Grey Heron "families" congregating together, flapping their wings and calling out, is a rare and magnificent spectacle.

PHOTO **张俊臣**

苍鹭

一种大型水边鸟类。主要以小型鱼类、虾、蛙和昆虫等为食。常长时间站立于浅水中耐心等待捕食机会。喜欢单独活动，飞行速度慢，颈缩成“S”形。

Grey Heron

The Grey Heron is a large waterbird that primarily feeds on small fish, shrimp, frogs, and insects. It often stands patiently in shallow water for extended periods, waiting for the perfect opportunity to catch its prey. Preferring solitary activities, it has a slow flight speed and tucks its neck into an "S" shape while in the air.

MOMENT 39

起起落落

黄河口河滩上，一个数量庞大的豆雁群。豆雁们有的在天空飞舞，有的在水里觅食、嬉戏。天地间，任逍遥。起起落落，是雁的一生，也是人的一生。

豆雁是大型雁类，成鸟体长 69–80 厘米，体重约3千克。喜群居，飞行时成有序的队列，有一字形、人字形等。飞行时双翼拍打用力，振翅频率高。

Ups and Downs

On the riverbank of the Yellow River Estuary, a large flock of Bean Goose can be seen. Some are flying in the sky, while others are foraging and playing in the water. In this realm, they roam freely. Their rises and falls mirror the journey of both goose and humans.

The Bean Goose is a large species of goose, measuring 69-80 centimeters in length and weighing around 3 kilograms. They prefer to live in groups, flying in orderly formations, such as lines or V-shapes. When in flight, they flap their wings vigorously and have a high flapping frequency.

PHOTO 薄华瑞

MOMENT 40

归去来兮

枝头上以夜鹭为主的一群鹭鸟。归来如风，飞去似箭。

夜鹭属中型涉禽，体长 46-60 厘米，是具有夜视能力的鸟类之一，喜夜晚外出捕食。这群典型的“夜猫子”在黄河口度过的是黑白颠倒的“鸟生”。

PHOTO 刘永民

Return of the Night Herons

A flock of Black-crowned Night Herons perches on the branches. They return like the wind and take flight like arrows.

The Black-crowned Night Heron is a medium-sized wading bird, measuring 46-60 centimeters in length, and is known for its ability to see well in low light. These typical "night owls" spend their lives in the Yellow River Estuary in a world turned upside down, where night is day and day is night.

MOMENT 41

在鸟的海洋中

这是多么摄人心魄的景象——成千上万的鸟儿在黄河口湿地的天空振翅翩飞，旋转又俯冲，掀起一层层鸟浪。

In the Ocean of Birds

What a breathtaking sight it is—thousands of birds flapping their wings in the sky above the Yellow River Estuary, swirling and plunging, creating waves of feathers that roll and crash.

PHOTO 李宪军

MOMENT 42

起飞

一群丹顶鹤从水中飞起，振翅冲上天空。

丹顶鹤是一种候鸟，在长时间的飞行迁徙过程中，一般是由一只比较强壮的丹顶鹤在前面开路，飞行时排成"人"字形，角度保持 120 度。

Takeoff

A group of Red -Crowned Cranes takes off from the water, their wings beating as they ascend into the sky.

The Red -Crowned Crane is a migratory bird. During long flights, a stronger crane typically leads the way, flying in a "人" formation with an angle of approximately 120 degrees.

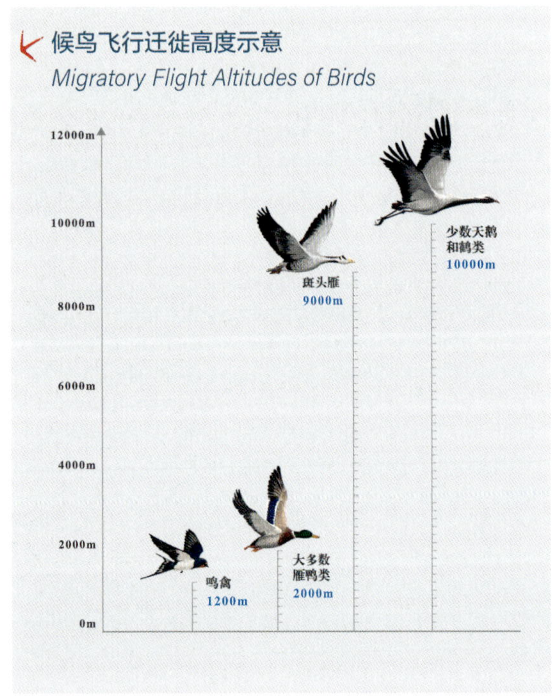

PHOTO 孙劲松

PHOTO 孙艳红

MOMENT 43

降落

一队白琵鹭，从空中依次降落到水面。

白琵鹭每年春秋两季栖息黄河口，偶有少量种群常年滞留。喜欢在芦苇沼泽、水塘、泥滩觅食，成群结队在浅水中缓慢前进。觅食时，头部往两边甩动，以鱼虾为食。白琵鹭飞翔时两翅鼓动较快，平均每分钟鼓动达 186 次左右。

Landing

A flock of Eurasian Spoonbills descends one after another to the water's surface.

Eurasian Spoonbills inhabit the Yellow River Estuary during the spring and autumn seasons, with some small populations occasionally staying year-round. They prefer to forage in reed marshes, ponds, and mudflats, moving slowly in groups through shallow waters. While foraging, they sway their heads side to side, feeding on fish and shrimp. When in flight, their wings beat rapidly, averaging about 186 beats per minute.

PHOTO 孙劲松

PHOTO 孙劲松

MOMENT 44

怒发冲冠

一只白鹭和一只黑翅长脚鹬在争地盘。“怒发冲冠”的白鹭显然在气势上更胜一筹。为了领地，鸟类之间的争夺战屡见不鲜。

Anger Rising

A Little Egret and a Black-winged Stilt engage in a territorial dispute. With feathers flared, the Little Egret clearly asserts its dominance. Such battles for territory are common among birds as they defend their spaces.

PHOTO 付建智

MOMENT 45

捷足先登

这张照片是摄影师在掩体里苦等了六个小时拍到的瞬间。两只蓝翡翠捕食回来，前后仅相差两秒，先到者抢占到了树桩的“使用权”，一脸得意。

First Come, First Served

This photo captures a moment the photographer waited six hours in a hide to witness. Two Black-capped Kingfishers return from their hunt, just two seconds apart. The first to arrive claims the "rights" to the perch, looking quite pleased with itself.

MOMENT 46

水上渔者

一只翠鸟水中捕食的精彩瞬间。翠鸟捕鱼技艺超群，如鱼叉一样，直插水面快如闪电。

这一幕完美贴合了小学语文课本中《翠鸟》里的描写：翠鸟蹬开苇秆，像箭一样飞过去，叼起小鱼，贴着水面往远处飞走了。只有苇秆还在摇晃，水波还在荡漾。

Fisher on Water

A stunning moment captured of a Kingfisher hunting in the water. The Kingfisher, like a skilled fisherman, strikes the surface with lightning speed, piercing the water like a harpoon.

This scene perfectly aligns with the description in the elementary school textbook story *The Kingfisher*: The Kingfisher launches off the reeds, flying like an arrow, snatching up a small fish and darting away just above the water's surface. Only the reeds continue to sway, and the water ripples linger.

PHOTO **冯展梅**

PHOTO 祝芳振

MOMENT 47

深夜食客

一只草鸮在午夜时分捕获了它最爱的食物。它眼睛内的视锥细胞密度是人眼的八倍，瞳孔很大，感光能力强，所以草鸮是妥妥的“夜间杀手”。它全身羽毛柔软蓬松，飞行的时候无声无息，能出其不意地捕杀食物。

草鸮，属鸮形目草鸮科草鸮属，俗名猴面鹰，属于夜行性猛禽，主要以鼠类、蛙类及小型鸟类为食，属国家二级保护动物。

Midnight Diner

A Eastern Grass Owl captures its favorite meal at the stroke of midnight. With a density of cone cells in its eyes eight times greater than that of humans, large pupils, and excellent light sensitivity, the Eastern Grass-owl is undoubtedly a "night predator." Its feathers are soft and fluffy, allowing it to fly silently and ambush its prey unexpectedly.

The Eastern Grass Owl belongs to the family Strigidae and the genus Asio, commonly known as the Monkey-faced Owl. It is a nocturnal raptor that primarily feeds on rodents, frogs, and small birds, and it is classified as a national second-class protected animal.

MOMENT 48

遭遇天敌来袭

水中嬉戏的一群骨顶鸡，突然被它们的天敌白腹鹞（又名泽鹞）袭击。鹞是生活在湿地中的一类猛禽，以小型动物为食。

骨顶鸡虽被称为“鸡”，但却是一种中型游禽，多成群结队，结伴而行。

Encounter with a Predator

A group of Eurasian Coots, playing in the water, were suddenly attacked by their predator, the Eastern Marsh Harrier (also known as the Marsh Harrier). Harriers are a type of raptor that lives in wetlands and preys on small animals.

Though Eurasian Coots are called "chickens", they are actually medium-sized waterbirds that tend to travel together in flocks.

PHOTO 孙劲松

MOMENT 49

水上芭蕾

一只白鹭在水面上展开双翅，如同一位芭蕾舞者沉浸在音乐中自我陶醉地跳舞。

白鹭属鹭科鸟类，具有嘴长、颈长、脚长特点，依赖于湿地环境栖息觅食。在黄河口，各种白鹭均可见，散布于黄河三角洲地区的滩涂、沼泽、河滩、稻田等处。

Water Ballet

A Little Egret spreads its wings above the water, resembling a ballet dancer lost in the music, gracefully performing its dance.

Little Egret belong to the heron family and are characterized by their long beaks, long necks, and long legs, relying on wetland environments for habitat and foraging. At the Yellow River Estuary, various species of Little Egrets can be seen scattered across the tidal flats, marshes, riverbanks, and rice fields of the Yellow River Estuary.

PHOTO 王娄

PHOTO 刘月良

MOMENT 50

空中伦巴

在芦苇荡上空纵情飞舞的一只草鹭，为那个隐身在芦苇丛中的“心仪之人”，跳起了一曲“爱情之舞”伦巴。

草鹭喜欢栖息在生长有大片芦苇和水生植物的水域，以小鱼、蛙、甲壳类、蜥蜴、蝗虫等为食。

Aerial Rumba

A Purple Heron dances joyfully in the sky above the reed marshes, performing a "love dance" rumba for its hidden admirer among the reeds.

Purple Herons prefer habitats with extensive reeds and aquatic plants, feeding on small fish, frogs, crustaceans, lizards, and grasshoppers.

MOMENT 51

诗意足印

一只丹顶鹤在黄河口的河滩上踱步，留下了一串串纤细而优雅的脚印，如同用精致的画笔在大地上轻轻勾勒出的图案。

Poetic Footprints

A Red -Crowned Crane strolls along the riverbank of the Yellow River, leaving behind a trail of delicate and elegant footprints, as if it were lightly sketching patterns on the earth with an exquisite brush.

PHOTO 任冠军

MOMENT 52

凝望远方

黄河口的滩涂上，一只鹭伫立在夕阳的霞光中，清澈的眼眸凝视着远方，有一种超脱尘世的宁静与淡然。

Gazing into the Distance

On the mudflats of The Yellow River Estuary,a Heron stands in the glow of the sunset,its clear eyes fixed on the horizon,embodying a tranquility and serenity that transcends the mundane.

PHOTO 郑旭东

PHOTO 王浩然

MOMENT 53

两情相悦

这是一个爱的瞬间。雌雄两只凤头䴙䴘似乎认定了对方就是自己命中注定的伴侣。

凤头䴙䴘，又名“浪里白”，黄河口常见鸟类，夏季在黄河口地区繁殖，生活在湿地的浅水区，以各种鱼类为食。凤头䴙䴘体长为 50 厘米以上，头后面直立着像凤头一样的两撮黑色翎毛，非常炫酷。

Mutual Affection

This is a moment of love. A pair of Great Crested Grebes seems to have found their destined companions in each other.

Commonly seen in The Yellow River Estuary, the Great Crested Grebe, also known as "White in Waves," breeds in shallow wetlands during the summer, feeding on various fish. With a body length of over 50 centimeters, it boasts two striking black plumes on the back of its head, resembling a regal crest.

PHOTO 黄建清

MOMENT 54

如果这就是爱情

两只太平鸟，站立在枝头，一果定情。自然界里出现的这一幕，或许就是人类认为的爱情最美的样子。

太平鸟为鸟纲太平鸟科的鸟类，属小型鸣禽，头顶有一细长呈簇状的羽冠，体态优美、鸣声清柔，以果实、种子、嫩芽等植物性食物为食。

If This Is Love

Two Bohemian Waxwings perch on a branch, a single fruit binding their affection. This scene in nature may embody the most beautiful form of love as perceived by humans.

The Bohemian Waxwing, a small passerine bird from the family Bombycillidae, features a slender, tufted crown atop its head. Graceful in form and soft in song, it feeds on fruits, seeds, and tender buds.

PHOTO **刘尊胜**

MOMENT 55

在一起

一对东方白鹳夫妇。东方白鹳 2 月开始产卵，2–7 月处于繁殖期。

Together

A pair of Oriental White Storks. Oriental White Stork begin laying eggs in February, with their breeding season extending from February to July.

进入繁殖期，东方白鹳开始搭窝筑巢，它们特别喜欢在高大乔木和电线杆上筑巢。

As they enter the breeding season, the Oriental White Storks begin to build nests. They particularly favor tall trees and utility poles for their nesting sites.

PHOTO **黄高潮**

PHOTO 刘永民

MOMENT 56

破壳而出

黑嘴鸥宝宝出壳了，祝你鸟生快乐！

一只黑嘴鸥的孵化期是 24-26 天，雌雄黑嘴鸥轮流负责孵化，每隔半小时换岗一次。

Hatching from the shell

The Saunders's Gull chick begins its journey—wishing you a joyful bird life!

The incubation period for a Saunders's Gull is 24-26 days, with both male and female taking turns to incubate, swapping every half hour.

MOMENT 57

世界那么大，我想去看看

黑嘴鸥宝宝从妈妈的翅膀下探出头来大声叫喊，它对这个崭新的世界充满了无限渴望。

**The world is so vast,
I want to see it all.**
The Saunders's Gull chick peeks out from under its mother's wings, calling out loudly, filled with endless longing for this brand-new world.

PHOTO **刘永民**

黑嘴鸥

属中型水鸟，喙是黑色的，头也是黑色的，所以被称为“头戴黑礼帽，身穿燕尾服的鸟类绅士”。黑嘴鸥对环境极其敏感，是一种典型的滨海滩涂鸟类。黑嘴鸥属于全球性易危物种，仅分布于东亚地区，目前估计世界种群数量 14400 只左右。黄河三角洲繁殖数量稳定在 10000 余只，是全球最大的黑嘴鸥繁殖地之一、中国三大黑嘴鸥繁殖地之一，被誉为“中国黑嘴鸥之乡”。

Saunders's Gull

Saunders's Gull are medium-sized waterbirds with distinctive black bills and heads, earning them the nickname "birds in black top hats and tailcoats."These gulls are highly sensitive to their environment and are typical inhabitants of coastal mudflats. As a globally vulnerable species, saunders's gulls are primarily found in East Asia, with an estimated world population of around 20,000 individuals. The Yellow River Estuary is home to a stable breeding population of over 10,000, making it one of the largest breeding grounds for Saunders's Gull in the world and one of China's top three breeding sites, earning it the title of "hometown of Saunders's Gull in China."

MOMENT 58

妈妈回来了

中华攀雀雏鸟终于盼来了出门觅食的妈妈，兴奋地大声呼喊。

中华攀雀是山雀科攀雀属的小型鸟类，被誉为“鸟类建筑大师”。中华攀雀的巢像一只浅灰色的毛绒“靴子”，密实、美观又舒适。鸟巢不需要任何支撑点，而是凌空系在树梢头，在空中悠悠地晃荡，颇有艺术气质。

Mom is back

The fledgling Chinese Penduline Tit eagerly calls out as its mother returns, ready to forage for food.

This small bird from the Paridae family is known as the “master builder” of the avian world. The nests of Chinese Penduline Tits resemble soft, gray felt “boots”—dense, attractive, and comfortable. Remarkably, these nests do not require any support; instead, they are suspended high in the treetops, gently swaying in the air, embodying an artistic flair.

PHOTO 孙崇伦

MOMENT 59

好好吃，快快长

这是一幅戴胜育雏的画面。小家伙最近食量大增，鸟妈妈每天需要多次捕食往返喂养。

戴胜，是戴胜科戴胜属鸟类，体长 30 厘米左右，遇到惊吓、紧张或兴奋时会张开头部羽毛。

Eat well, grow fast

This scene captures a Eurasian Hoopoe nurturing its young. The little ones have recently developed a hearty appetite, requiring the mother bird to make multiple foraging trips each day to feed them.

The Eurasian Hoopoe, belonging to the family Upupidae and the genus Upupa, measures about 30 centimeters in length. When startled, anxious, or excited, it will flare its head feathers.

PHOTO **黄高潮**

MOMENT 60

跟着妈妈学走路

反嘴鹬宝宝跟着妈妈蹒跚学步。妈妈教得认真，宝宝学得有样。

反嘴鹬是一种腿特别长的水鸟，体长 38-45 厘米，生活在湿地和靠近海湾的湖里。背部有醒目的黑色和白色标志，腹部灰白色，喙向上弯曲，主要吃水里的昆虫、小鱼、贝类和两栖动物。

Learning to Walk with Mom

The baby Pied Avocet wobbles along behind its mother, who teaches with great care while the little one tries to imitate her.

The Pied Avocet is a long-legged water bird measuring 38-45 centimeters in length, found in wetlands and lakes near coastal bays. It features distinctive black and white markings on its back, a grayish-white belly, and an upward-curving bill, primarily feeding on aquatic insects, small fish, shellfish, and amphibians.

PHOTO 刘永民

PHOTO 赵文昌

MOMENT 61

妈妈，我会跑了

小黑嘴鸥挺胸抬头，迈着小碎步，一摇一摆地跑向前方。

黑嘴鸥宝宝从破壳到摇摇摆摆地走来走去，只需要 1 个小时，30 天之后便可在空中翱翔。

Mom, I Can Run Now

The Saunders's Gull puffs out its chest, taking tiny, wobbly steps as it runs forward.

From hatching to walking unsteadily, a Saunders's Gull chick only takes an hour. In just 30 days, it will be soaring through the skies.

MOMENT 62

走红毯

绿头鸭妈妈率众宝宝列队出行，走在碱蓬草铺就的“红地毯”上。

绿头鸭是现代养殖家鸭的野生祖先之一，是一种分布广泛的雁形目鸭科鸭属水禽，又被称为野鸭、大麻鸭、大绿头、大红腿鸭等，以其雄性鲜艳的绿色头部和白色颈环而闻名。

Walking the Red Carpet

The Mallard mother leads her line of ducklings, parading along the "red carpet" of seepweed.

The Mallard is one of the wild ancestors of modern domestic ducks, a widely distributed waterfowl belonging to the Anatidae family. Known by names such as Wild Duck, Big Greenhead, and Big Red-legged Duck, it is distinguished by the male's striking green head and white neck ring.

PHOTO 刘月良

MOMENT 63

夕阳漫步

黑翅长脚鹬夫妻带着宝贝们在夕阳下漫步。

黑翅长脚鹬是一种中型涉禽，体长约 37 厘米，拥有一双修长而优雅的大长腿，行走时步态轻盈稳健，被誉为鸟界的“超模”。以软体动物、昆虫等为食。性胆小而机警，栖息于湖泊、沼泽浅水地带。

Sunset Stroll

The Black-winged Stilt couple strolls with their little ones under the glowing sunset.

The Black-winged Stilt is a medium-sized wading bird, measuring about 37 centimeters in length, known for its long and elegant legs. With a light and steady gait, it is often regarded as the "supermodel" of the bird world. Feeding on mollusks and insects, this species is shy and vigilant, preferring to inhabit shallow waters of lakes and marshes.

PHOTO 崔乡选

PHOTO 刘涛

MOMENT 64

瞧这一家子

枝头上的一个长耳鸮家庭，相貌与表情相似度极高，如同复制粘贴。

长耳鸮，别名长耳猫头鹰、夜猫子，栖息于山地森林或平原树林中。主要以鼠类和昆虫为食。 属于国家二级保护动物。

Look at This Family

A family of Long-eared Owls perches on a branch, their strikingly similar appearances and expressions resembling a copy-and-paste job.

The Long-eared Owl, also known as theLong-eared Cat Owl or Night Owl, inhabits mountainous forests and plain woodlands. Its diet primarily consists of rodents and insects. This species is classified as a national second-level protected animal.

PHOTO **张东河**

MOMENT 65

父爱如山

高大强壮的草鹭爸爸守护着草鹭妈妈和草鹭宝宝们，尽显“一家之主”风范。这是展现动物界“父爱如山”的生动一幕。

Father's love is like a mountain

The tall, strong Purple Heron father stands protectively over the mother and their chicks, embodying the role of "head of the family". This is a vivid scene showcasing "Father's love is like a mountain" in the animal kingdom.

MOMENT 66

中国好声音

红喉歌鸲与红胁绣眼鸟对唱。鸟儿的歌喉是大自然最美妙的声音。

China's Voice

The Siberian Rubythroat and Chestnut-flanked White-eye sing together in harmony. The birds' melodies are among the most beautiful sounds in nature.

PHOTO 付建智

MOMENT 67

同唱一首歌

小白鹭在妈妈的指导下学唱。童声合唱，歌声嘹亮。

白鹭种群繁殖期为 4-9 月，雌雄鸟共同筑巢，产卵数 3-5 枚，孵化期 19-21 天，育雏期近 1 个月。

One Song Together

The little Egret learns to sing under its mother's guidance. Their chorus rings out, bright and clear.

The breeding season for Egrets lasts from April to September, during which both male and female build the nest together, laying 3 to 5 eggs. The incubation period lasts 19 to 21 days, and the chicks are cared for for nearly a month.

PHOTO 马厚全

MOMENT 68

练功夫

小黄斑苇鳽停在芦苇枝头，努力伸展着身体，似乎是在修炼武功，来日一鸣惊人。

黄斑苇鳽是一种中型涉禽，常沿沼泽、苇塘飞翔，性机警，遇到敌害时迅速隐藏到植物丛中一动不动，以小鱼虾、昆虫为食。

Practicing Kung Fu

The little Yellow Bittern perches on a reed branch, stretching its body as if training in martial arts, preparing to astonish the world one day.

The Yellow Bittern is a medium-sized wading bird that often flits along marshes and reed beds. Cautious by nature, it quickly hides among the vegetation when danger approaches, feeding on small fish, shrimp, and insects.

PHOTO 付建智

PHOTO 全再明

MOMENT 69

攀高枝

棕头鸦雀攀在一根芦苇的高处眺望远方，似在憧憬着自己的未来。

棕头鸦雀是一种小型鸟类，为较常见留鸟，体色整体呈棕粉褐色。常栖息于芦苇地、灌丛和林缘等地带，以各类昆虫等为食，也吃果实和种子等。

Reaching for Higher Branches

The Vinous-throated Parrotbill perches high on a reed, gazing into the distance, as if dreaming of its future.

This small bird is a common resident, with an overall plumage of brown and pinkish-brown hues. It typically inhabits reed beds, shrubs, and forest edges, feeding on various insects as well as fruits and seeds.

PHOTO 刘庆堂

MOMENT 70

好奇看世界

一对纵纹腹小鸮充满好奇地四处张望。

纵纹腹小鸮，鸱鸮科小鸮属的一种鸟类，是体型很小的猫头鹰。无耳羽簇，头顶平，眼亮黄而长凝不动。属国家二级保护动物。

Curiously Observing the World

A pair of Little Owls Gazes around with wide-eyed curiosity.

These Little Owls belonging to the genus Athene, are known for their lack of ear tufts and flat crowns, with bright yellow eyes that remain fixed in stillness. They are classified as a national second-class protected species.

MOMENT 71

享受慢生活

夕阳的余晖中，一个数量约 20 只的卷羽鹈鹕鸟群栖息于芦苇沼泽中。卷羽鹈鹕在东亚迁徙种群种仅 100 余只，黄河三角洲有超过 50 只（最多时 84 只），为卷羽鹈鹕最大的迁徙停歇地。

Enjoying Slow Living

In the afterglow of the sunset, a group of about 20 Dalmatian Pelicans rests in the reed marshes. In East Asia, the migratory population of Dalmatian Pelicans is only around 100 individuals, with over 50 (peaking at 84) spotted in the Yellow River Estuary, making it the largest stopover site for this species during migration.

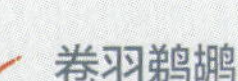

卷羽鹈鹕

一种大型的白色水鸟，体长 1.6–1.8 米，翼展长 3 米左右，多生活在沼泽及浅水湖，以鱼类、甲壳类、软体动物、两栖动物等为食。卷羽鹈鹕在世界自然保护联盟濒临灭绝物种危急清单中被列为“易危”物种，全球卷羽鹈鹕的数量在 10000–20000 只左右。

Dalmatian Pelican

The Dalmatian Pelican is a large white waterbird, measuring between 1.6 to 1.8 meters in length, with a wingspan of up to 3 meters. According to the International Union for Conservation of Nature (IUCN), the Dalmatian Pelican is classified as "Vulnerable," with an estimated global population of around 10,000 to 20,000 individuals.

PHOTO 马子光

MOMENT 72

春色伊人

春天的黄河口，一只白鹭静静地守望在水草丰沛的湿地上。

所谓伊人，在水一方。

Spring's Beauty

In the springtime at the Yellow River Estuary, a solitary Little Egret watches over the lush wetlands.

As the saying goes, "She stands by the water's edge."

PHOTO 郑旭东

MOMENT 73 仲夏鹭羽

一群牛背鹭在黄河口夏日的花海中翩翩起舞。

牛背鹭是唯一不食鱼而以昆虫为主食的鹭类。其与家畜尤其是水牛形成了依附关系，常跟随水牛捕食，也常在牛背上歇息，因此得名。每年 4 月初到 4 月中旬迁到北方繁殖地，9 月末至 10 月初迁离繁殖地到南方越冬。

PHOTO 李刚

Midsummer Egret Feathers

A flock of Eastern Cattle Egrets dances gracefully in the sea of flowers at the Yellow River Estuary on a summer day.

The Eastern Cattle Egret is the only heron species that primarily feeds on insects rather than fish. It forms a symbiotic relationship with livestock, especially water buffalo, often following them to hunt and resting on their backs, which is how it gets its name. Every year, from early to mid-April, they migrate to northern breeding grounds, departing by late September to early October for their southern wintering sites.

MOMENT 74

鹭影秋静

白琵鹭在黄绿相间的芦苇荡上空飞翔，构成了元好问笔下的“秋色图”：鹭影兼秋静，蝉声带晚凉。陂长留积水，川阔尽斜阳。

Egrets in Autumn Stillness

The Eurasian Spoonbill soar over the green and yellow reeds, creating a scene reminiscent of Yuan Haowen's "Autumn Colors": shadows of egrets blend with the tranquil autumn, cicadas sing of the evening chill. Water lingers in the ponds, while the vast river stretches beneath the slanting sun.

PHOTO **于绍源**

雀舞迎冬

瑞雪纷飞，树麻雀为了进食，纷纷登枝抢食。

树麻雀别名麻雀、家贼，是雀形目雀科麻雀属小型鸟类。性活泼，常集群活动，一般在地面、草丛及灌丛中觅食。

Sparrows Dance to Welcome Winter

As snowflakes flutter down, Eurasian Tree Sparrow eagerly perch on branches, vying for food.

Also known as House Sparrows or "Thieving Sparrows," these small birds belong to the Passeridae family. They are lively and often gather in flocks, typically foraging on the ground, in grasslands, and among shrubs.

PHOTO **李宪军**

MOMENT 76

捉迷藏

一只野兔在红色碱蓬草里探头探脑，好像在和谁玩捉迷藏。黄河三角洲野兔数量众多，分布广。

Hide-and-seek

A wild rabbit peeks out from the red seepweed, as if playing a game of hide-and-seek. The Yellow River Estuary is home to various animals, with a significant population of wild rabbits widely distributed throughout the area.

PHOTO 刘月良

PHOTO 刘月良

MOMENT 77

彩羽飞天

一只雉鸡展开它彩色的羽翼，振翅起飞。雉鸡，又名环颈雉，黄河三角洲留鸟，栖于不同高度的开阔林地、灌木丛等多种生境。

Colorful Wings in Flight

A Common Pheasant spreads its vibrant plumage and takes to the skies. Also known as the Ring-necked Common Pheasant, this bird is a resident of the Yellow River Estuary, inhabiting various environments, including open woodlands and shrublands at different elevations.

PHOTO 潘伟峰

MOMENT 78

天使的翅膀

阳光映照水面，将白鹭染成金色。白鹭展开双翅，如同一位天使降临人间。

Angel's Wings

The sunlight reflects off the water's surface, bathing the Little Egret in a golden hue. As it spreads its wings, it resembles an angel descending to Earth.

PHOTO 吴兆英

MOMENT 79

隐藏人物

停在树上的一只长耳鸮，毛色与树干融于一体，像一位隐藏得很深的“大人物”。

Hidden Figure

Perched on a tree, a Long-eared Owl blends seamlessly with the bark, resembling a deeply concealed "figure of importance."

PHOTO **赵英丽**

MOMENT 80

亲子合影

黑嘴鸥妈妈和它的两个宝宝在傍晚的霞光里留下一张温馨甜蜜的亲子合影。

Family Portrait

In the evening glow, a Saunders's Gull mother poses for a heartwarming family portrait with her two chicks, capturing a moment of sweet tenderness.

PHOTO 罗雅君

MOMENT 81

群鸟逐“鹳”

群鸟在黄昏中飞翔追逐的剪影。黄河口上空，一场激烈的飞鸟竞速赛正在进行，东方白鹳遥遥领先，后面的鸟紧追不舍，上演了群鸟争冠的精彩一幕。

Chasing the Stork

In the twilight sky, a silhouette of birds soars in pursuit of one another. Above the Yellow River Estuary, an intense bird race unfolds, with the Oriental White Stork leading the pack. The chasing birds trail closely behind, creating a thrilling scene of competition among the flock.

MOMENT 82

秋水长天

两只东方白鹳在镜子一样的水面上闲淡游走，秋水共长天一色。

Autumn Waters and Endless Skies

Two Oriental White Storks leisurely glide across the mirror-like surface of the water, where the autumn waters blend seamlessly with the endless sky.

PHOTO 李宪军

MOMENT 83

月中鹭

这个美得如诗如画的瞬间，任何语言的描述都是多余。

Herons in the Moonlight

In this moment, as beautiful as a poem or painting, any description feels superfluous.

PHOTO 陈勇

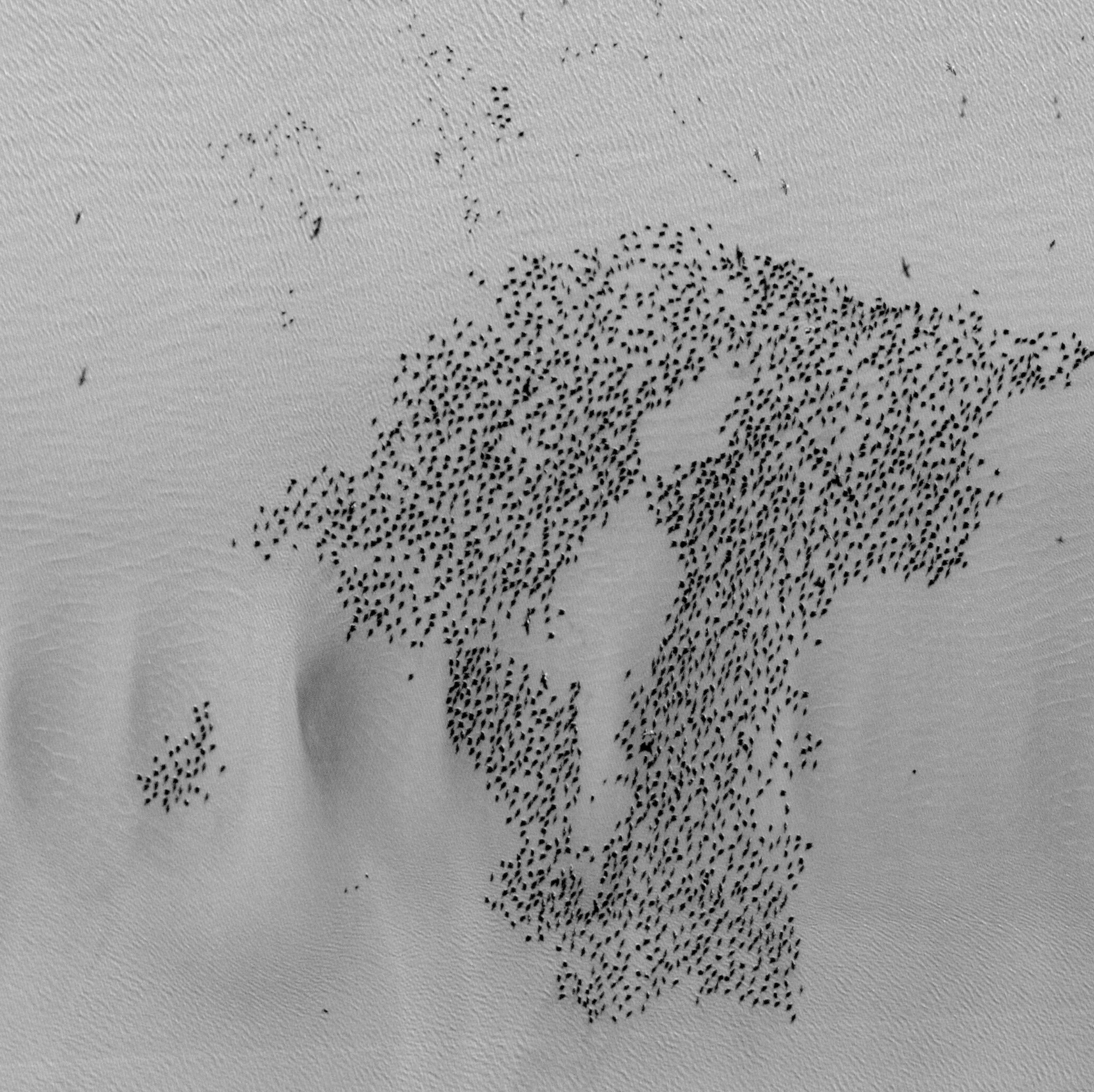

The World of Birds

From a drone's perspective, thousands of birds rest upon the Yellow River's banks, resembling a vast map of the world.

This is the realm of birds. They are the true masters of this world.

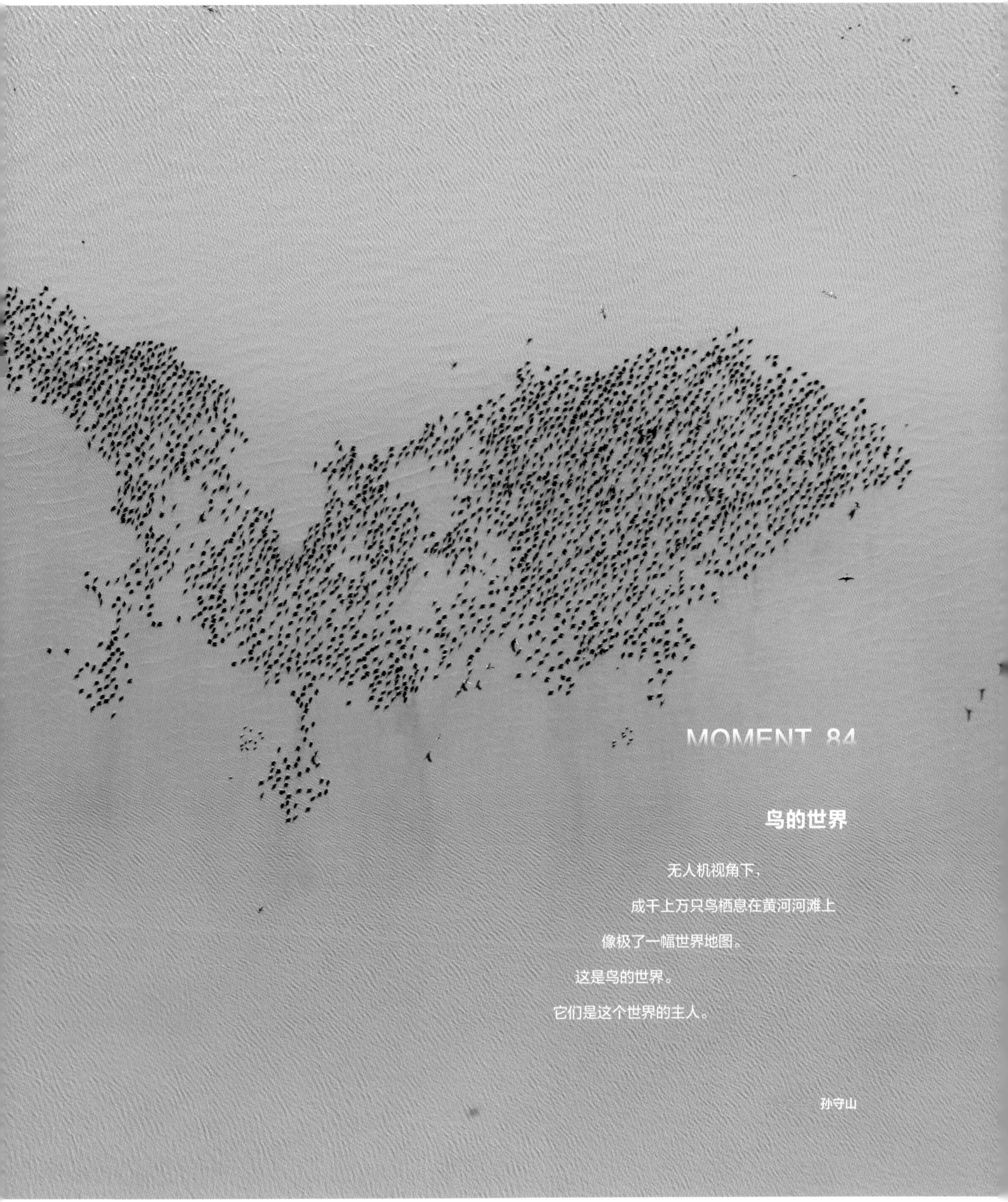

MOMENT 84

鸟的世界

无人机视角下，

成千上万只鸟栖息在黄河河滩上

像极了一幅世界地图。

这是鸟的世界。

它们是这个世界的主人。

孙守山

Humanistic Concern

The high-level protection and quality development of the Yellow River Estuary National Park is an earnest directive from General Secretary Xi Jinping during his inspection on October 20, 2021. As guardians of this land, we build "The Bridge of Life" for the natural wonders of the Yellow River Estuary using human wisdom, scientific conservation, technological support, and systematic management. In shared empathy, we create a beautiful tapestry of harmonious coexistence between humanity and nature.

人文守望

高水平保护、高质量建设黄河口国家公园，是2021年10月20日习近平总书记视察黄河口时的殷切嘱托。作为这方土地的守护人，我们以人类的智慧、科学的保护、科技的支撑、系统的管控为黄河口的自然万物搭建“生命之桥”，在共情陪伴中绘出一幅人与自然和谐共生的美好画卷。

100个决定性瞬间

100 Decisive Moments

黄河口国家公园

The Yellow River Estuary National Park

MOMENT 85

跨河越海的巡检

黄河口国家公园海域面积达 2147 平方公里，范围广、面积大。国家公园管理机构不断完善巡检监控，并综合应用卫星遥感、无人机、智能视频、海洋浮标在线监测系统等手段，建立全天候快速响应的天空、地面、海洋一体化监测体系，建设智慧国家公园。

Inspections Across Rivers and Seas

The sea area of the Yellow River Estuary National Park covers 2,147 square kilometers, making it extensive and vast. The management agency of the national park continuously enhances inspection and monitoring capabilities by integrating satellite remote sensing, drones, smart video, and ocean buoy online monitoring systems. This establishes a comprehensive monitoring system for rapid, all-weather response across sky, ground, and sea, contributing to the creation of a smart national park.

PHOTO **赵文昌**

MOMENT 86

野外监测

一位黄河口的科研人员在进行野外监测。

黄河口与中国科学院、中国环科院等 30 余家国家级科研机构合作，成立了黄河口湿地野外科学观测研究站、黄河三角洲滨海湿地生态试验站等 11 家野外监测和科研教学平台，联合开展关键物种研究、湿地修复模式、外来有害物种防治等科研攻关，形成了 20 余项可复制推广的科研成果。

Field Monitoring

A researcher at the Yellow River Estuary is conducting field monitoring.

The Yellow River Estuary collaborates with over 30 national research institutions, including the Chinese Academy of Sciences and the Chinese Academy of Environmental Sciences, to establish 11 field monitoring and research platforms such as the Yellow River Wetland Field Science Observation Research Station and the Yellow River Delta Coastal Wetland Ecological Experiment Station. These platforms work together on key species research, wetland restoration models, and the control of invasive species, resulting in more than 20 replicable and promotable scientific achievements.

PHOTO 张目

MOMENT 87

人工巢里幸福安家

觅食的东方白鹳妈妈飞回它建在高架塔上的家，两只白鹳宝宝正在家里等着它的归来。

东方白鹳的繁殖期一般从 2 月持续到 7 月，黄河三角洲缺少适合东方白鹳筑巢的高大乔木。黄河口管理机构针对东方白鹳繁殖特点和分布情况，设计出了适合它们的人工招引巢，稳稳托起了东方白鹳幸福的家。

Happily Settled in an Artificial Nest

A foraging Oriental White Stork mother returns to her home built on a high platform, where her two chicks eagerly await her arrival.

The breeding season for the Oriental White Stork lasts from February to July. However, the Yellow River Delta lacks suitable tall trees for the oriental white stork to nest. In response to the breeding characteristics and distribution of the Oriental White Stork, the Yellow River Estuary management agency has designed artificial nesting structures tailored for them, providing a stable home for their happiness.

利用视频监控实时掌握东方白鹳繁殖过程，为开展科学研究、掌握野外生存第一手资料提供支持。

Using video surveillance to monitor the breeding process of the Oriental White Stork in real-time, providing support for scientific research and acquiring firsthand information on their survival in the wild.

PHOTO **杨廷奎**

PHOTO 赵英丽

MOMENT 88

水润湿地

黄河口高质量推进国家公园建设，恢复黄河三角洲湿地水循环体系，近三年生态补水均超过 1.5 亿立方米，不断修复湿地结构和功能，不断提高湿地生物多样性。

Watering the Wetlands

The high-quality development of the Yellow River Estuary National Park has focused on restoring the water circulation system of the Yellow River Delta wetlands. In the past three years, ecological replenishment has exceeded 1.5 billion cubic meters, continuously repairing the structure and function of the wetlands and enhancing their biodiversity.

PHOTO 丁洪安

MOMENT 89

水相连

水系相连、陆海相通的湿地，被落日镀上金光。黄河口修复了“河—陆—滩—海”连通体系，连通水系 115 公里，疏通潮沟 76 公里，恢复了黄河与海洋的水文连通。

Linked by Waterways

The wetlands, with interlinked waterways and a seamless connection between land and sea, are bathed in golden light by the setting sun. The Yellow River Estuary has restored the connectivity of the "river-land-tide-sea" system, linking 115 kilometers of waterways and clearing 76 kilometers of tidal channels, thereby restoring the hydrological connection between the Yellow River and the ocean.

Gazing at the Starry Moonlit Night

The Viewing Tower at the Yellow River Estuary stands against the starry sky, resembling a painting from Van Gogh's Impressionist era.

Designed by the Tongji University Architectural Design Institute, this unique structure features a suspended design supported by four core tubes, covering a total area of 7,616 square meters. It serves multiple functions, including dining, waiting for boats, and sightseeing, making it a super-modern, multifunctional building.From the high observation platform, visitors can gaze at the Yellow River flowing into the sea, establishing it as one of the "Top Ten New Landmarks of Qilu Culture."

The Yellow River Estuary continually enhances its cultural infrastructure, using the power of culture to deepen humanity's understanding of nature and strengthen the harmonious coexistence between people and the environment.

MOMENT 90

远望星月夜

黄河口远望楼在星空的映衬下，如同梵高的印象派油画。

远望楼由同济大学建筑设计研究院设计，建筑风格独特，由四个核心筒支撑起的悬空结构，总面积 7616 平方米，是集餐饮、候船、观光等多功能于一体的综合性超现代建筑。站在高高的观景平台上，可以远眺黄龙入海，是“十大齐鲁文化新地标”之一。

黄河口不断完善人文建设，以文化的影响力加强人对自然的了解，密切人与自然的和谐共生。

PHOTO **赵英丽**

黄河口鸟类博物馆：全国规模最大的鸟类专题博物馆之一

Yellow River Estuary Bird Museum:One of the Largest Bird Themed Museums in the Country

PHOTO **清风明月（视觉中国 供图）**

天然柳林木栈道：在万亩柳林上观鸟赏鱼

Natural Willow Forest Boardwalk: A Place to Birdwatch and Fish in the Vast Willow Woods

PHOTO **张目**

MOMENT 91

感受生态之美

一对情侣手牵手置身黄河口秋日碱蓬草染红的滩涂上，如同踏上了爱的“红地毯”。

依托黄河口国家公园建立的黄河口生态旅游区是国家 5A 级旅游景区，如今已成为人们感受生态之美，体验自然神奇的重要旅游目的地。

Experiencing the Ecological Beauty

A couple walks hand in hand on the red-tinted mudflats of the Yellow River Estuary, surrounded by autumn's seepweed, as if stepping onto a "red carpet" of love.

The Yellow River Estuary Ecotourism Area, established under the Yellow River Estuary National Park, is a national 5A-level tourist attraction and has become an important destination for people to experience the beauty of ecology and the wonders of nature.

PHOTO 杨廷奎

PHOTO 赵英丽

MOMENT 92

走近自然课堂

与灰雁的第一次亲密接触，让这个小男孩无比兴奋。无数孩子来到黄河口国家公园鸟类科普乐园研学，这里已经成为他们探索自然奥秘的生动课堂。

依托黄河三角洲国家级自然保护区独特的自然环境，黄河口正在努力打造成黄河口生态文明研学基地。

Approaching the Classroom of Nature

The first close encounter with a Greylag Goose fills the little boy with immense excitement. Countless children visit the Yellow River Estuary Bird Science Park for educational exploration, turning it into a vibrant classroom for uncovering the mysteries of nature.

Leveraging the unique natural environment of the Yellow River Delta National Nature Reserve, the Estuary is working diligently to develop into an ecological civilization research and learning base.

PHOTO 张目

黄河口国家公园鸟类科普乐园是集救护、繁育、科普为一体的大型开放式鸟类乐园。在这里可以近距离观赏到东方白鹳、丹顶鹤、大天鹅、蓑羽鹤等珍稀鸟类，并能通过鸟类介绍牌了解鸟类的习性，学习鸟类知识。

The Yellow River Estuary Bird Science Park is a large, open bird sanctuary that combines rescue, breeding, and education. Here, visitors can closely observe rare bird species such as the Oriental White Stork, Red-crowned Crane, Whooper Swan, and the Siberian Crane. Informational panels provide insights into the birds' habits, allowing guests to learn more about avian knowledge.

MOMENT 93

执法巡察

黄河口国家公园积极探索多部门协同执法途径，不断提升执法效能，为构建陆海统筹型综合执法体系奠定了良好基础。

Law Enforcement Patrols and Inspections

The Yellow River Estuary National Park Management Bureau's Comprehensive Law Enforcement Detachment conducts patrols in the water areas under its jurisdiction. The park actively explores collaborative law enforcement approaches involving multiple departments, continuously enhancing enforcement effectiveness and laying a solid foundation for establishing a coordinated terrestrial and marine comprehensive law enforcement system.

PHOTO 张目

MOMENT 94

大豆生于野

黄河口野大豆保育区里的一株野大豆长势喜人。野大豆是我国第一批重点保护野生植物，属国家二级保护植物，具有重要科研价值。

黄河三角洲是目前我国仅有大面积拥有野大豆的地方，为加强野大豆资源的保护，黄河口划定野大豆保育区 7.46 万亩，保障野大豆自然生长演替功能，推动生物多样性保护。

Wild Soybeans in the Wild Land

A wild soybean plant thrives in the Yellow River Estuary Wild Soybean Conservation Area. Wild soybeans are among China's first group of key protected wild plants and are classified as a national secondary protected species, holding significant scientific research value. The Yellow River Delta is currently the only region in China with a large area of wild soybeans. To strengthen the conservation of wild soybean resources, the Yellow River Estuary has designated a 74,600-acre conservation area, ensuring the natural growth and succession functions of wild soybeans and promoting biodiversity protection.

PHOTO 张目

PHOTO 张目

MOMENT 95

小雪重生

在黄河口国家公园鸟类科普乐园，天鹅“小雪”展开双翅，似乎想拥抱它的驯养师李建，向他表达自己无尽的谢意。

2007 年，有村民发现了受伤严重的小雪，救护人员的细心呵护让它获得了重生。至今，小雪已在黄河口国家公园鸟类科普乐园里幸福生活了近 20 年。

A New Life for Little Snow

At the Yellow River Estuary Bird Science Park in Shandong, the swan named "Little Snow" spreads its wings, seemingly wanting to embrace its caretaker, Li Jian, to express endless gratitude.

In 2007, villagers discovered Little Snow severely injured. The dedicated care from rescue personnel gave it a second chance at life. Now, Little Snow has happily lived in the Yellow River Estuary Bird Science Park for nearly 20 years.

PHOTO 赵文昌

MOMENT 96

安居美家

黑嘴鸥妈妈在专门为她打造的栖息地里安心养育宝宝。

黄河口开展东方白鹳、黑嘴鸥等关键物种栖息地的保护工作，自2010年开始，根据黑嘴鸥的繁殖需求，划定3200亩潮间带作为核心保护区域，通过建设围堤、隔坝，实施封闭式管理及分区域管理，为其孵育后代提供场所，为其量身打造安全舒适的“繁衍之家”。

A Safe and Beautiful Home

The Saunders's Gull mother nurtures her chicks peacefully in a habitat specially created for her.

Since 2010, the Yellow River Estuary has focused on protecting critical habitats for key species such as the Saunders's Gull and the Oriental White Stork. In response to the breeding needs of the Saunders's Gull, 3,200 acres of intertidal zones have been designated as core protection areas. Measures such as constructing levees and barriers, implementing closed management, and managing different zones provide a safe and comfortable "home for reproduction" for these birds to raise their young.

MOMENT 97

观鸟亭上

在观鸟亭上，看群鸟起落，听鸟鸣齐喧，蔚为壮观。鸟群在黄河口的天幕上，尽情挥洒着生命的活力，数万双羽翅切割空气所发出的声响如万马奔腾。“鸟云”似流动的画，像跳跃的诗，身临其境，你一定会被大自然的神奇所折服和震撼。

At the Birdwatching Pavilion

From the birdwatching pavilion, one can witness the spectacle of flocks of birds rising and falling, accompanied by a symphony of chirps and calls. The birds dance across the sky at the Yellow River Estuary, exuberantly expressing the vitality of life. The sound of thousands of wings slicing through the air echoes like a galloping herd of horses. The "cloud of birds" resembles a flowing painting, a poem in motion; being immersed in this scene, you will surely be captivated and awed by the wonders of nature.

PHOTO **刘月良**

PHOTO 赵文昌

MOMENT 98

生命之光

当我们用最朴素的方式融入荒野，不打扰、不惊动，这难道不是人与自然最和谐美好的一幕吗？

The Light of Life

When we blend into the wilderness in the simplest of ways, without disturbing or startling the surroundings, isn't this the most harmonious and beautiful scene between humans and nature?

MOMENT 99

定格瞬间

一位摄影师在黄河口拍摄空中的飞鸟。黄河口的美景吸引了全国各地的摄影家前来采风。中国摄影家协会已经将这里作为“中国摄影创作基地”。

Captured Moments

A photographer captures flying birds in the skies over the Yellow River Estuary. The stunning scenery attracts photographers from across the country, and the China Photographers Association has designated this location as a "China Photography Creation Base."

PHOTO 赵英丽

MOMENT 100

你与黄河口的瞬间

The Moment you are with the Yellow River Estuary

此处请存放你与黄河口国家公园邂逅的珍贵瞬间

This is the place to showcase your precious moments of encounters with the Yellow River Estuary National Park.

PHOTO 周广学

让我们记住这 **100 个决定性瞬间，**

这是黄河口之所以成为中国最美荒野的 100 个理由；
让我们珍视自然的每一次生命的悸动，
这是荒野馈赠给人类的伟大礼物……

这里有光与黑暗、生与死；
这里有几乎是永恒的时间；
这里有能量与生物进化，
创造出多产与勇力、适应与创制、
信息与生物进化、对抗与顺应；
这里有肌肉与脂肪、神经与汗水、
规律与形式、结构与过程、
美丽与聪明、和谐与庄严、灾祸与荣耀。

Let's remember these 100 decisive moments,
which are 100 reasons why the Yellow River Estuary
is one of China's most beautiful wildernesses.
Let us cherish every heartbeat of life in nature,
a magnificent gift from the wild to humanity...
Here, there is light and darkness, life and death;
here, time feels almost eternal; here, there is energy and biological evolution,
creating abundance and courage, adaptation and innovation,
information and biological evolution, resistance and compliance;
here, there are muscles and fat, nerves and sweat,
patterns and forms, structure and processes,
beauty and intelligence, harmony and solemnity, disaster and glory.

荒野是一个有投射与选择能力的系统，
编织出了一个内容丰富的故事。
荒野是我们在现象世界中
能经验到的生命最原初的基础，
也是生命最原初的动力。

The wilderness is a system with the ability to project and select,
weaving a rich tapestry of stories.
It is the most primal foundation of life
we can experience in the realm of phenomena,
and it is also the most fundamental drive of existence.

Appendix

黄河口观鸟时间表

The Yellow River Estuary Birdwatching Schedule

黄河口一年四季皆可观鸟，不同季节，鸟类族群不同。从鸟种来看，春秋迁徙季节鸟类族群最丰富，3 月中下旬及 11 月上中旬是理想的观鸟时间。从观察难度来看，夏、冬鸟类易观察，3–7 月是观东方白鹳的好季节，冬季观雁类、鸭类、天鹅类较为理想。

Birdwatching is possible year-round at the Yellow River Estuary, with different bird populations present in each season. The richest variety of species occurs during the spring and autumn migration seasons. Optimal times for birdwatching are late March to mid-April and early November. In terms of observation difficulty, summer and winter birds are easier to spot. The best time to observe the Oriental White Stork is from March to July, while winter is ideal for watching geese, ducks, and swans.

一月 January

越冬雁类、鸭类、天鹅类、鹤类（主要是丹顶鹤、灰鹤）、大鸨（有的年度）。分布于芦苇沼泽、近海滩涂、农田、草地、水库、河流等环境。

Wintering geese, ducks, swans, and cranes (mainly Red-crowned Cranes and Common Cranes), as well as Great Bustards (in some years), can be found in reed marshes, coastal mudflats, farmland, grasslands, reservoirs, and rivers.

二月 February

同一月。中下旬，在此越冬的雁类、鸭类开始北迁，但南方的雁类、鸭类同时迁来，种类组成变化很大。

During the same month, from mid to late February, the geese and ducks that have been wintering here begin their northward migration. However, at the same time, geese and ducks from the south migrate in, resulting in significant changes in the species composition.

三月 March

中下旬，越冬的雁类、鸭类、天鹅类迁离，北迁的鹤类、鸥类、鹳类、鸻类、鹬类、鹭类等抵达。分布于芦苇沼泽、近海滩涂、水库、河流等环境。

In mid to late March, the wintering geese, ducks, and swans depart, while migratory cranes, gulls, storks, plovers, snipes, and herons arrive. These birds can be found in environments such as reed marshes, coastal mudflats, reservoirs, and rivers.

四月 April

北迁的鸻类、鹬类、雀形目鸟类抵达，同时在此繁殖的鹳类、鹭类、鸥类等开始筑巢、求偶、交配、产卵。分布于芦苇沼泽、近海滩涂、盐池、养殖池、河流、潮沟等环境。

Migratory plovers, snipes, and passerine birds arrive, while breeding storks, herons, and gulls begin nesting, courtship, mating, and laying eggs. These birds are distributed in environments such as reed marshes, coastal mudflats, salt ponds, aquaculture ponds, rivers, and tidal channels.

五月 May

上旬，北迁的鸻类、鹬类、雀形目鸟类迁离，猛禽类迁徙过境；中旬以后，鸟类种群相对稳定，主要为夏候鸟。芦苇沼泽、近海滩涂、园林苗圃、盐池等环境较为理想。

In early May, migratory plovers, snipes, and passerine birds depart, while raptors pass through during migration. After mid-May, the bird populations stabilize, primarily consisting of summer visitors. Ideal environments include reed marshes, coastal mudflats, nursery gardens, and salt ponds.

六月至八月 June to August

鸟类孵化、育雏及雏鸟生长期。主要种类有鸥类、鸻类、鹬类、鹭类等。这个季节鸟类种类较少、数量分散，由于植被生长茂盛，不易观察。芦苇沼泽、近海滩涂、盐池等环境较为理想。

This period involves bird incubation, chick-rearing, and the growth phase of young birds. The main species include gulls, plovers, snipes, and herons. During this season, bird diversity is lower, and numbers are scattered. Due to lush vegetation, observation becomes more challenging. Ideal environments include reed marshes, coastal mudflats, and salt ponds.

九月 September

雀形目鸟类开始抵达，鸥类、鸻类、鹬类部分抵达。芦苇沼泽、近海滩涂、盐池等环境较为理想。

Passerine birds begin to arrive, with some gulls, plovers, and snipes also present. Ideal environments for observation include reed marshes, coastal mudflats, and salt ponds.

十月 October

中下旬，雁鸭类、鹤类、鸥类、鹳类等抵达。分布于芦苇沼泽、近海滩涂、水库、河流、盐池、潮沟等环境。

In mid to late October, geese, ducks, cranes, gulls, and storks arrive. They are found in reed marshes, coastal mudflats, reservoirs, rivers, salt ponds, and tidal channels.

十一月 November

上中旬，南迁的大部分水禽类抵达；同时，越冬的雁类、鸭类、天鹅类、鹤类抵达。这个季节是鸟类种类最为丰富的时期。分布于芦苇沼泽、近海滩涂、水库、河流、盐池、潮沟等环境。

In early to mid-November, most waterfowl migrate south, along with wintering geese, ducks, swans, and cranes. This season is the most diverse for bird species. They are found in reed marshes, coastal mudflats, reservoirs, rivers, salt ponds, and tidal channels.

十二月 December

越冬的雁类、鸭类、天鹅类、鹤类等。分布于芦苇沼泽、近海滩涂、农田、草地、水库等环境。

Wintering geese, ducks, swans, and cranes are present. They can be found in reed marshes, coastal mudflats, farmland, grasslands, and reservoirs.

PHOTO 赵瑞祥

As long as I live,
I will listen to the songs of the wind, the birds, and the waterfalls;
I will understand the language of rocks, floods, storms, and landslides.
I will befriend the wilderness and glaciers,
Drawing as close to the heart of nature as I can.
Indeed, I do this:
I wander among the rocks,
Stroll through the forests,
Trek along the streams.
Whenever I encounter a new plant,
I will sit beside it for a minute
Or a day,
Trying to make friends and listen to what it wants to say.
I ask the pebbles where they come from and where they are going.
When night falls, I camp on the spot.
I am unhurried and unflustered,
Leisurely as the trees and stars.
This is true freedom—
A beautiful, attainable immortality.

— John Muir,
The founder of the modern environmental movement
and the father of the national parks in the United States.

群鸟飞翔在河海交汇、蓝黄分界线的上空，摆出了一个心形的图案，仿佛是对黄河口这片生态家园表达深切的爱意。

A flock of birds soars above the confluence of river and sea, forming a heart-shaped pattern in the sky, as if to express their deep affection for the ecological haven of the Yellow River Estuary.

PHOTO 丁洪安

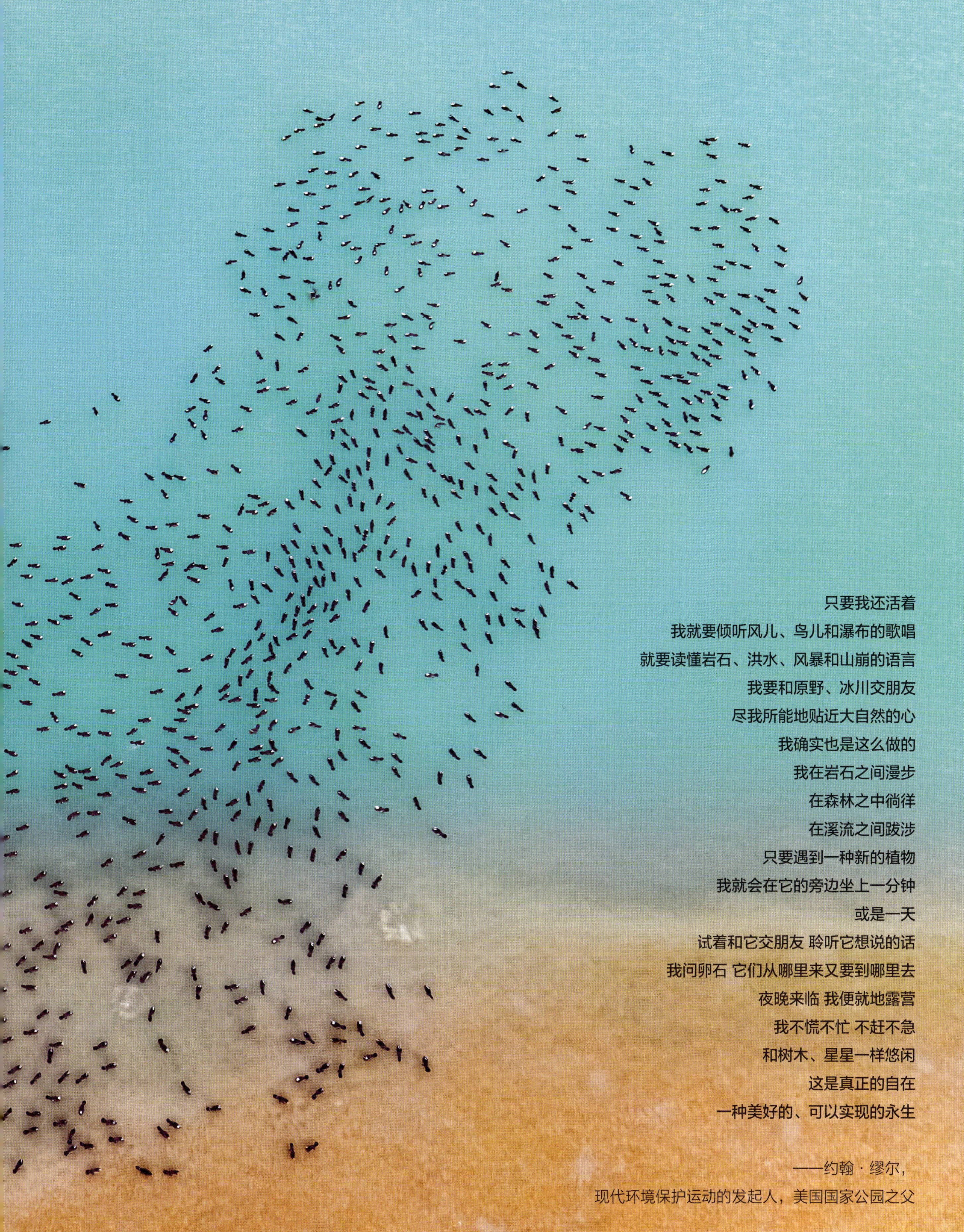

只要我还活着
我就要倾听风儿、鸟儿和瀑布的歌唱
就要读懂岩石、洪水、风暴和山崩的语言
我要和原野、冰川交朋友
尽我所能地贴近大自然的心
我确实也是这么做的
我在岩石之间漫步
在森林之中徜徉
在溪流之间跋涉
只要遇到一种新的植物
我就会在它的旁边坐上一分钟
或是一天
试着和它交朋友 聆听它想说的话
我问卵石 它们从哪里来又要到哪里去
夜晚来临 我便就地露营
我不慌不忙 不赶不急
和树木、星星一样悠闲
这是真正的自在
一种美好的、可以实现的永生

——约翰·缪尔，
现代环境保护运动的发起人，美国国家公园之父

HUANGHEKOU GUOJIA GONGYUAN 100 GE JUEDINGXING SHUNJIAN

黄河口国家公园 100 个决定性瞬间

出版发行 山东画报出版社